# STOREY'S BARN GUIDE TO SHEEP

Storey Publishing

*The mission of Storey Publishing is to serve our customers by publishing practical information that encourages personal independence in harmony with the environment.*

Edited by Deborah Burns, Sarah Guare, and Dale Perkins

Art direction and cover design by Vicky Vaughn

Text design and production by Jessica Armstrong

Cover photograph by © Buck Kelly / Getty Images

Interior photographs by © age fotostock America, Inc., iv and 28; © Animals Animals/ Earth Scenes/ Steven David Miller, 10; © Mark E. Gibson, 74; © Peter Dean/ Grant Heilman Photography, Inc., 36 and 88

Illustrations by © Colin Hayes Illustrator, Inc.

Illustrations reviewed by Elizabeth Smith and Dr. Ann Wells, DVM

Indexed by Christine R. Lindemer, Boston Road Communications

Copyright © 2006 by Storey Publishing, LLC

Portions of this book have been adapted from *Storey's Guide to Raising Sheep* by Paula Simmons and Carol Ekarius.

All rights reserved. No part of this book may be reproduced without written permission from the publisher, except by a reviewer who may quote brief passages or reproduce illustrations in a review with appropriate credits; nor may any part of this book be reproduced, stored in a retrieval system, or transmitted in any form or by any means — electronic, mechanical, photocopying, recording, or other — without written permission from the publisher.

The information in this book is true and complete to the best of our knowledge. All recommendations are made without guarantee on the part of the author or Storey Publishing. The author and publisher disclaim any liability in connection with the use of this information. For additional information please contact Storey Publishing, 210 MASS MoCA Way, North Adams, MA 01247.

Storey books are available for special premium and promotional uses and for customized editions. For further information, please call 1-800-793-9396.

Printed in the United States by CJK

10 9 8 7 6 5 4 3 2 1

LIBRARY OF CONGRESS CATALOGING-IN-PUBLICATION DATA

Storey's barn guide to sheep.
p. cm.
Includes index.
ISBN-10: 1-58017-849-9;
ISBN-13: 978-1-58017-849-5 (pbk. : alk. paper)
1. Sheep. I. Title: Barn guide to sheep. II. Title: Guide to sheep.

SF375.S86 2006
636.3—dc22 2006005933

# CONTENTS

CHAPTER ONE

# Getting Acquainted

# TEN TIPS FOR HEALTH & HAPPINESS

**EVERYONE HAS HIS OR HER OWN SET OF TIPS,** but here are some good ones from veteran sheep farmers:

1. **ESTABLISH REGULAR FEEDING TIMES.** Feedings are usually given once in the morning and once in the evening.

2. **STICK TO THE REGULAR DIET.** Add the new feed and reduce the old feed over a period of several days to avoid making the sheep sick when you do need to change feed.

3. **PROVIDE PLENTY OF FRESH WATER.** Water is an important part of your sheep's good health, and they will drink more if it's fresh.

4. **FEED BALANCED RATIONS.** Base your sheep's diet on feed analysis and nutritional requirements. If your sheep are getting their nutritional requirements through quality hay or pasture, don't feed grain.

5. **PRACTICE GOOD HOOF CARE.** Trim hooves once or twice a year.

6. **CONSTRUCT GOOD FENCES.** Everyone will be happier if the sheep remain where you would like them to be.

7. **COMPOST THAT MANURE.** Well-maintained pens will keep your sheep healthier, provide a good working environment, reduce flies, and help the garden grow.

8. **THINK LIKE A SHEEP.** Minimize stress. Herd and work around your sheep quietly with consistent gentle handling.

9. **PROVIDE SHADE.** An open-sided shed, shade trees, or a canopy roof can keep sheep cool in the summer.

10. **PROVIDE PREDATOR PROTECTION.** Coyote-proof fencing, llamas, donkeys, and guard dogs can protect against predators.

# Basic Sheep Anatomy

**FAMILIARIZE YOURSELF WITH BASIC SHEEP ANATOMY** to communicate more easily with the veterinarian about your sheep.

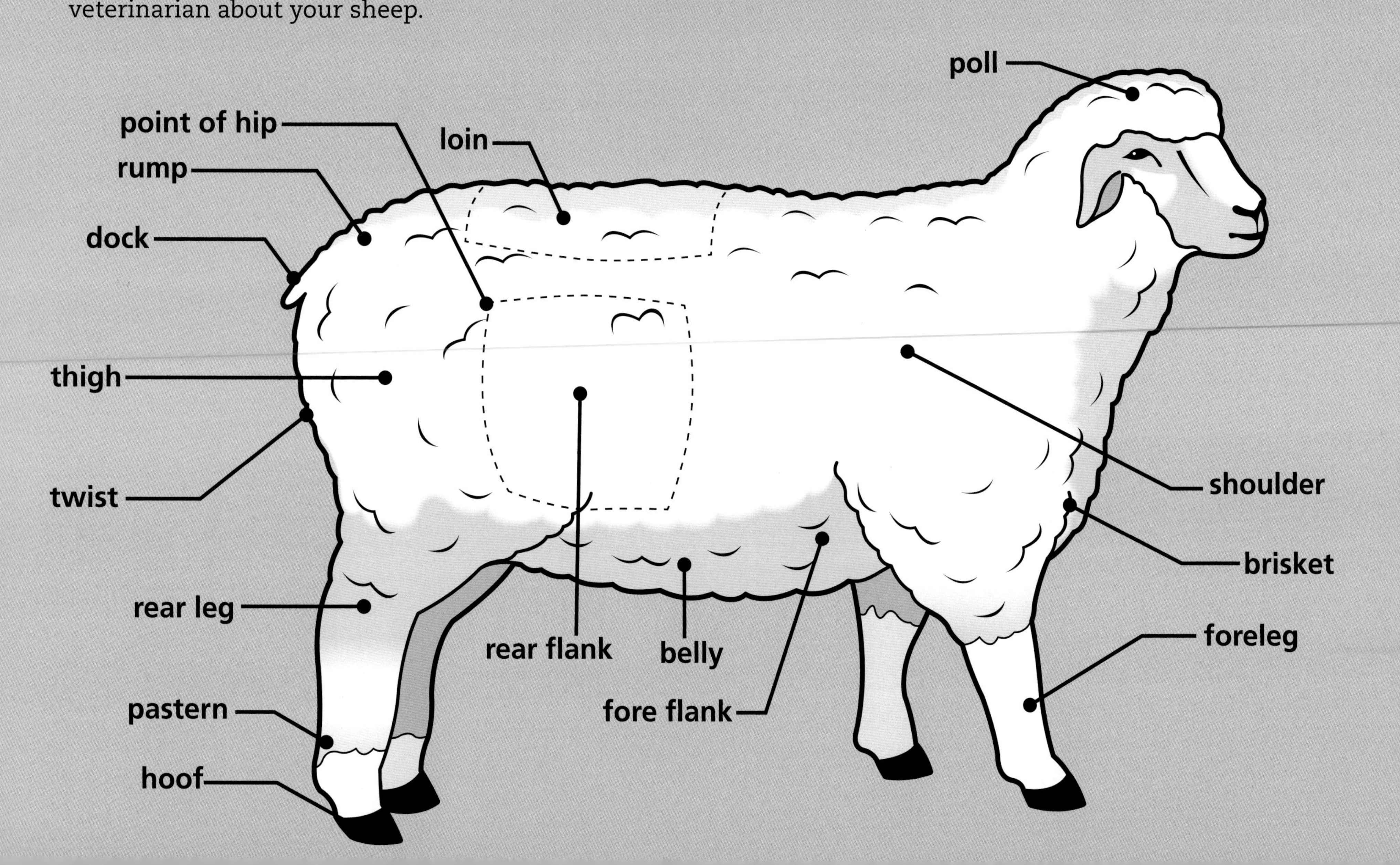

# ESTIMATING WEIGHT

**IF YOU DON'T HAVE A SCALE,** you can estimate an animal's weight using the technique described here. If the sheep is in full fleece, part the wool so that the measurement is accurate.

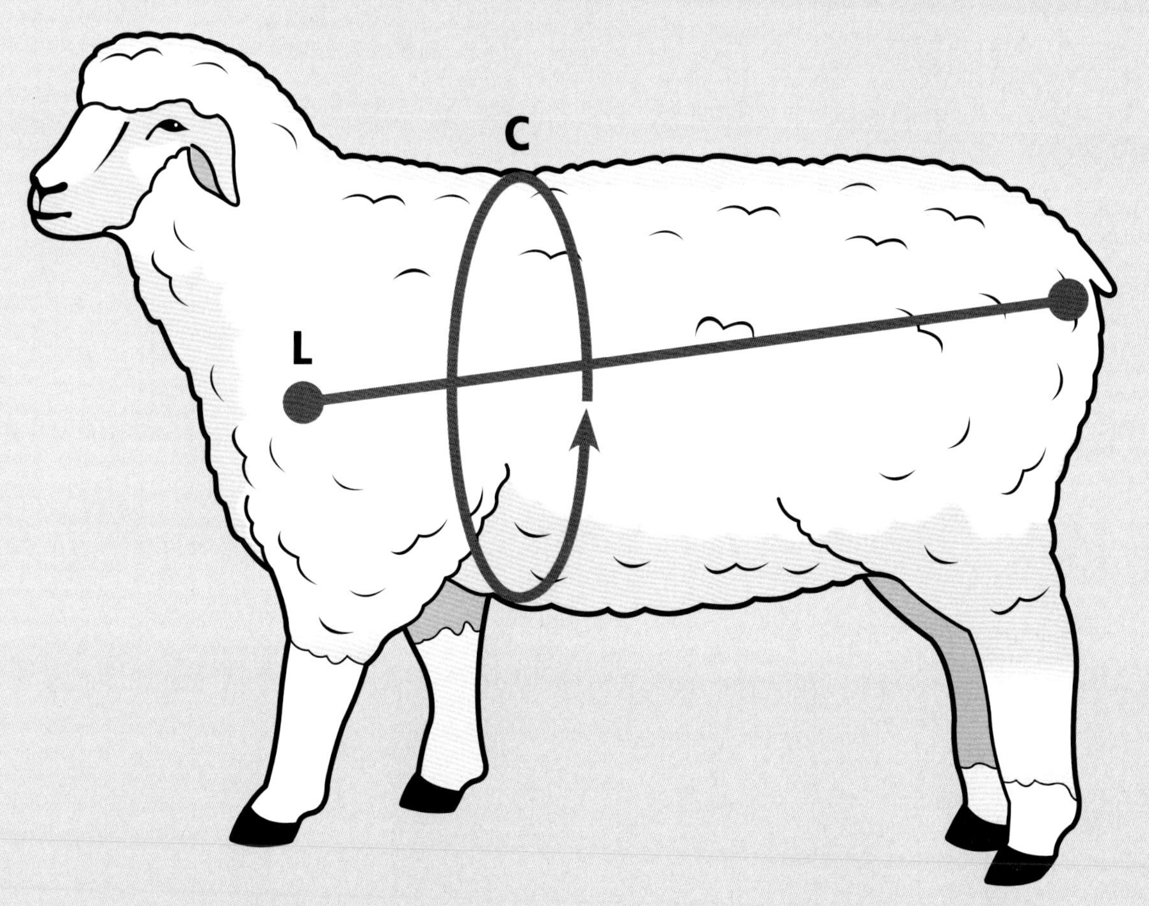


**1 MEASURE CIRCUMFERENCE (C)**

With a tape measure, measure all the way around the sheep's body just behind the front legs.

**2 MEASURE LENGTH (L)**

Next, measure the length of the body from the point of the shoulder to the point of the rump.

**3 CALCULATE WEIGHT**

Now, multiply C × C × L and divide by 300.

$$\text{Weight} = \frac{C \times C \times L}{300}$$

# Determining Age

**TO SOME EXTENT,** you can determine the age of sheep by their teeth. Sheep have no teeth in their upper jaw. In the front bottom jaw, they have four pairs of incisor teeth that change with age. The grinding action of chewing wears sheep's teeth. As the incisors wear down, the amount of tooth below the gum line is gradually pushed out to help compensate for the wear (see diagram below).

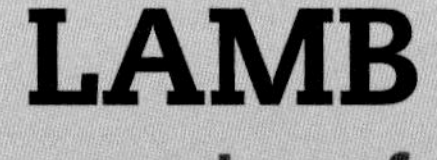

## LAMB

A lamb's four pairs of incisors are small baby teeth that fall out to make way for permanent teeth.

## YEARLING

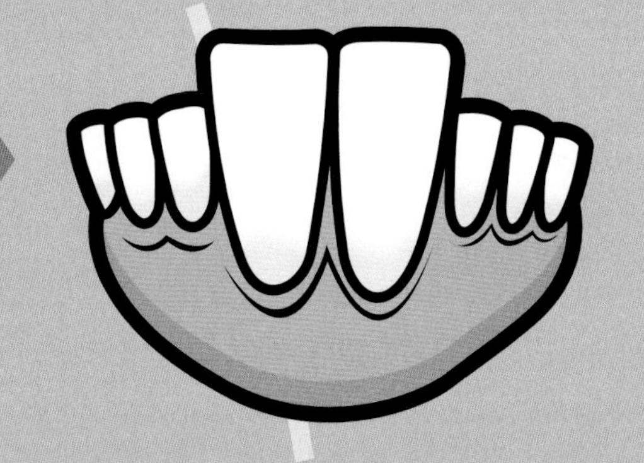

At about 1 year of age, the center pair fall out and the first pair of larger, permanent incisors appear.

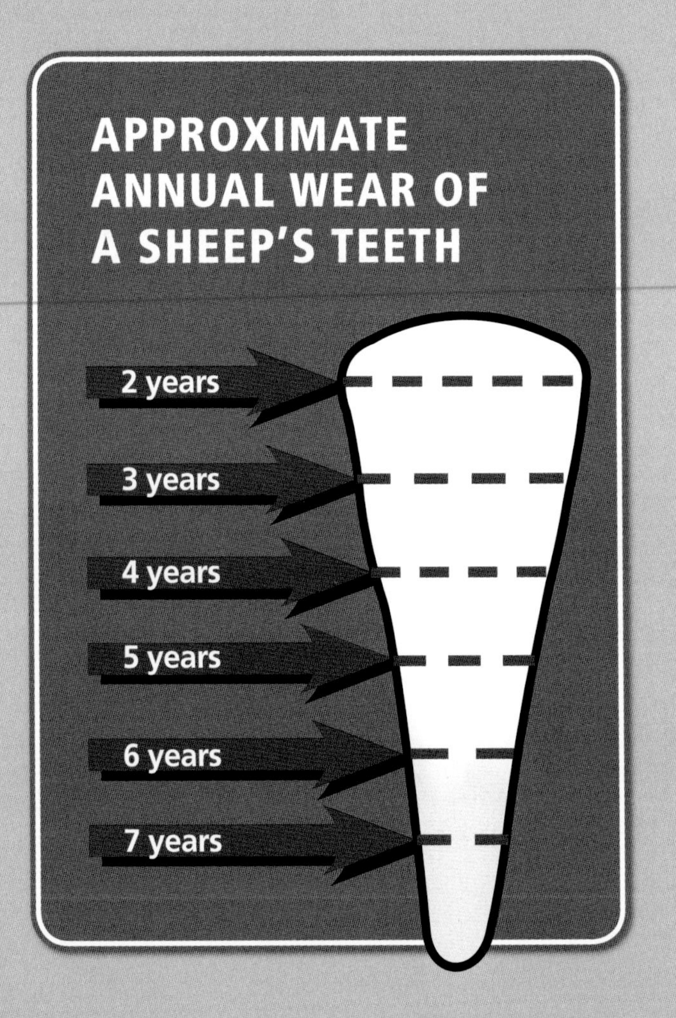

## YEARLING TO 4 YEARS

For each year until the sheep is 4 years old, it loses one pair of baby teeth and gains one pair of permanent teeth.

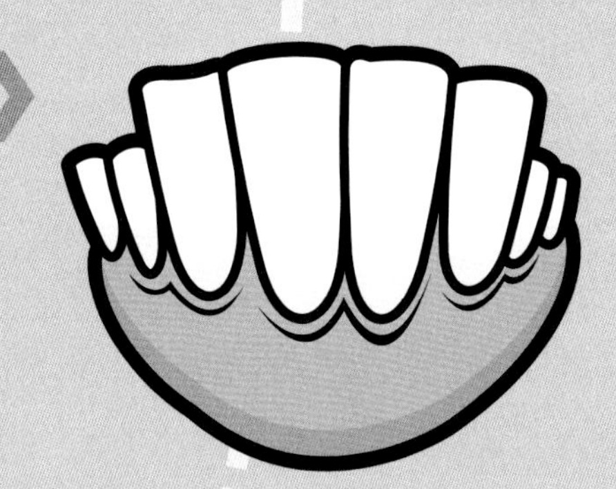

## TEETH TERMS

- SOLID MOUTH. Up to 4 years old; all adult teeth in place.
- SPREADERS. Older than 4 years; narrower parts of teeth have moved up from the gum.
- BROKEN MOUTH. Some teeth missing. Ewes may have one or two lambing seasons left.
- GUMMERS. All front teeth are lost.

## 4 YEARS

By now, a sheep has the fullest, most complete set of teeth it will have in its lifetime.

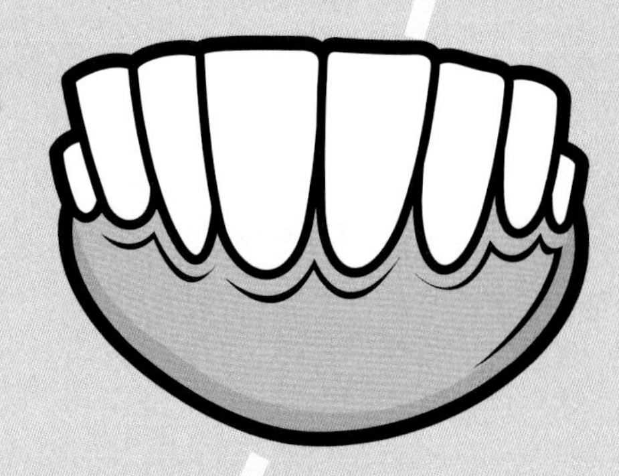

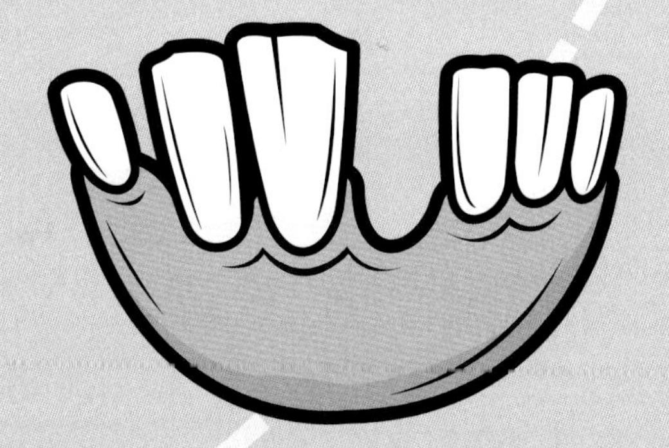

## OLDER THAN 4 YEARS

After a sheep turns 4 years old, it may have gaps and begin losing permanent teeth.

# Shepherd's Calendar

## ABOUT THIS CALENDAR

- The checklist will help you remember what needs to be done to care for your sheep throughout the year.
- The page numbers tell where in this guide you can find more information about each task.
- The calendar starts with May because that is when most new shepherds begin their flocks.

## MAY

- [ ] Check your fences before bringing home new sheep.
- [ ] Check your pasture for toxic plants (page 32).
- [ ] Give new sheep hay before turning them on to lusher pasture than they are used to.
- [ ] Check for keds and treat your sheep if you see even one ked.

## JUNE

- [ ] Practice pasture rotation (page 34).
- [ ] Provide plenty of water, salt, and mineral/vitamin mix.
- [ ] Watch for fly-strikes (clusters of tiny eggs) and maggots.
- [ ] Keep rear ends trimmed to discourage flies.

## JULY

- [ ] Provide shade for your sheep in hot weather.
- [ ] Provide 1 to 2 gallons of cool, fresh water daily for each adult sheep.
- [ ] Check the feet of limping sheep; trim hooves if necessary (page 16).
- [ ] Check for parasites by fecal count and deworm if necessary. Record date of deworming.

**If you are planning for fall breeding:**

- [ ] Put your ram in a shaded area near the ewe.
- [ ] Shear the ram's scrotum to keep him cool.
- [ ] Start flushing the ewe before you plan to bring in the ram for breeding.
- [ ] Body-condition ewes (page 20).

## AUGUST

- Continue to provide shade.
- Consider necessary parasite prevention and deworm if necessary.
- Consider purchasing hay and/or grain for winter. Store grain in rodent-proof containers that sheep cannot get into.

**If you are planning January lambing:**

- For early lambs, keep the ram in a shady place during the day and bring him in at night.
- Maintain ram condition; give him ¼–½ pound of grain daily if necessary.
- Keep the ram with your ewe for at least 6 weeks so they have two chances or more for mating. A marking harness on the ram will indicate that mating is occurring.
- Mark on your calendar the date you turned in the ram with the ewe, so you will know the earliest date to expect the lambing. Also mark the date if you observe any mating take place.
- Body-condition ewes (page 20).

## SEPTEMBER

- Locate a place to purchase your winter supply of hay.
- Have water, salt, and mineral/vitamin supplement always available to sheep.
- Make a list of repairs needed on shelter, fencing, and equipment, and start the repairs before cold weather.
- Continue parasite control. If you are deworming, be sure to read the label for any precautions about timing before slaughter.
- Get sizable lambs to market. Shear them first, or ask for their pelts if you have a good wool market.
- Clean out the place where you plan to store winter hay; save the manure you collect and spread it on the vegetable garden.
- If you have apples, feed a few (but not too many at once) to sheep. Set aside some wind-falls for winter.

**If you are breeding your sheep:**

- For late lambs, flush the ewe (add grain to diet) and then bring in the ram.
- Reduce grain gradually after flushing.
- Record dates when you see breeding take place (page 92).

## OCTOBER

- Clean out sheep sheds and barn and spread sheep manure on garden.
- Have your winter hay and feed plan ready.
- Clean out and check your waterers; winterize the faucets.

**If you are breeding your sheep:**

- If your ram harness indicates no activity, purchase or borrow another ram. The ewes of some breeds may quit cycling by late January.
- Check over your lambing supplies (page 41). Order supplies by mail if necessary.
- Make lamb jugs.
- Body-condition ewes (page 20).

# Shepherd's Calendar (continued)

## November

- ☐ **Keep grain in rodent-proof containers and take steps to get rid of rodents.**
- ☐ **Order antibiotics and store them in a refrigerator for emergencies.**
- ☐ **If any of your sheep are limping, check hooves and trim or treat if necessary (page 16).**

**If you have bred your sheep:**

- ☐ Check your lambing supplies (page 41). Order ear tags, if you plan to use them (page 65).
- ☐ Separate the ram from pregnant ewes so he doesn't injure them.
- ☐ Body-condition sheep (page 20) and adjust ration accordingly (page 31). If the ram is run-down, feed him well.
- ☐ Add a small amount of stock molasses to the pregnant ewes' drinking water.
- ☐ Make lamb jugs if you haven't already.

## December

- ☐ **Put molasses into ewes' drinking water to keep the water from freezing and to add sugar to the ewes' diet.**

**If you have bred your sheep:**

- ☐ Crotch the ewes to prepare them for lambing. Remove dirty tags from their udder and legs.
- ☐ Begin checking the udder of each pregnant ewe; if a ewe's udder is hard and lumpy, she may have no milk, and you should be prepared to bottle-feed her lamb.
- ☐ Four weeks before lambing time, begin feeding ewes ¼–½ pound of grain daily with their hay.
- ☐ If a ewe is listless, she may have pregnancy toxemia; call your veterinarian.
- ☐ Consider calcium supplement for your pregnant ewes if their diet is low in calcium.
- ☐ Body-condition sheep (page 20) to evaluate if they need supplemental nutrition.

## January

**If you have bred your sheep:**

- ☐ Watch pregnant ewes carefully for signs of labor. Now may be the time to start nightly lamb checks.
- ☐ Be sure pregnant ewes are getting exercise.
- ☐ If a ewe refuses to eat, she may have pregnancy toxemia or be about to lamb.
- ☐ Crotch ewes if you haven't done so.
- ☐ Add molasses to pregnant ewes' drinking water (1 cup per gallon).
- ☐ Have clean, dry lambing pens ready.
- ☐ After a lambing, put the ewe and the lamb into the jug. Strip the ewe's teats and "snip and dip" the lamb's umbilical cord (page 64). Make sure newborns get colostrum within 2 hours of birth. Give the ewe warm molasses water.
- ☐ Dock lambs' tails when they are 2 to 3 days old (page 22) and castrate male lambs when they are about 10 days old (page 24). Ear-tag newborns with lamb tags (page 65) and record birthing information.

## FEBRUARY

- [ ] Make sure salt is available.

**If you have lambs:**

- [ ] Watch lambs to be sure that they are having normal bowel movements.
- [ ] Watch twin lambs to be sure that one isn't growing more rapidly than the other; if that happens, supplement the feed for the slower-growing lamb.
- [ ] If a ewe loses a lamb, you may want to try grafting an orphan lamb (page 70).
- [ ] Ear-tag lambs (page 65).
- [ ] Check each ewe's feet and trim and treat them, if necessary, before taking sheep out of the pen (page 16).
- [ ] Give plenty of fresh, clean water.
- [ ] Continue feeding grain to nursing ewes.
- [ ] Prepare the lamb creep (an enclosure that allows lambs to enter for supplemental feeding, but prohibits older animals from entering).

## MARCH

- [ ] Check sheep for parasites and if necessary deworm all sheep and restock deworming supplies.
- [ ] Start shearing if weather permits (page 75). Keep mothers with their lambs as much as possible to avoid confusion. Do not shear wet sheep. Keep fleeces clean.
- [ ] At shearing time or 10 days later, treat for keds if they are present or have been a problem with your flock.
- [ ] Trim hooves at shearing time (page 16).
- [ ] Clean out the barn or shed and put old hay and manure on the vegetable garden or spread over an area of the pasture that needs to be fertilized.

## APRIL

- [ ] Place your feeder on well-drained ground to avoid hoof trouble.

**If you have lambs:**

- [ ] Keep fresh water, salt, and feed in the lamb creep.
- [ ] If your management system is set up for it, let in lambs before turning ewes onto new pasture, so they can get the best of the grass.
- [ ] Have your vet take a fecal count on lambs when they weigh about 40 pounds and deworm if neccessary.
- [ ] You may want to wean them at this time.
- [ ] Before lambs are weaned, decrease the ewes' grain ration and feed them only hay to reduce milk production.

### APPROXIMATE EQUIVALENTS

1 pound (lb) ≈ ½ kilogram

1 gallon ≈ 4 liters

CHAPTER TWO

# Basic Care

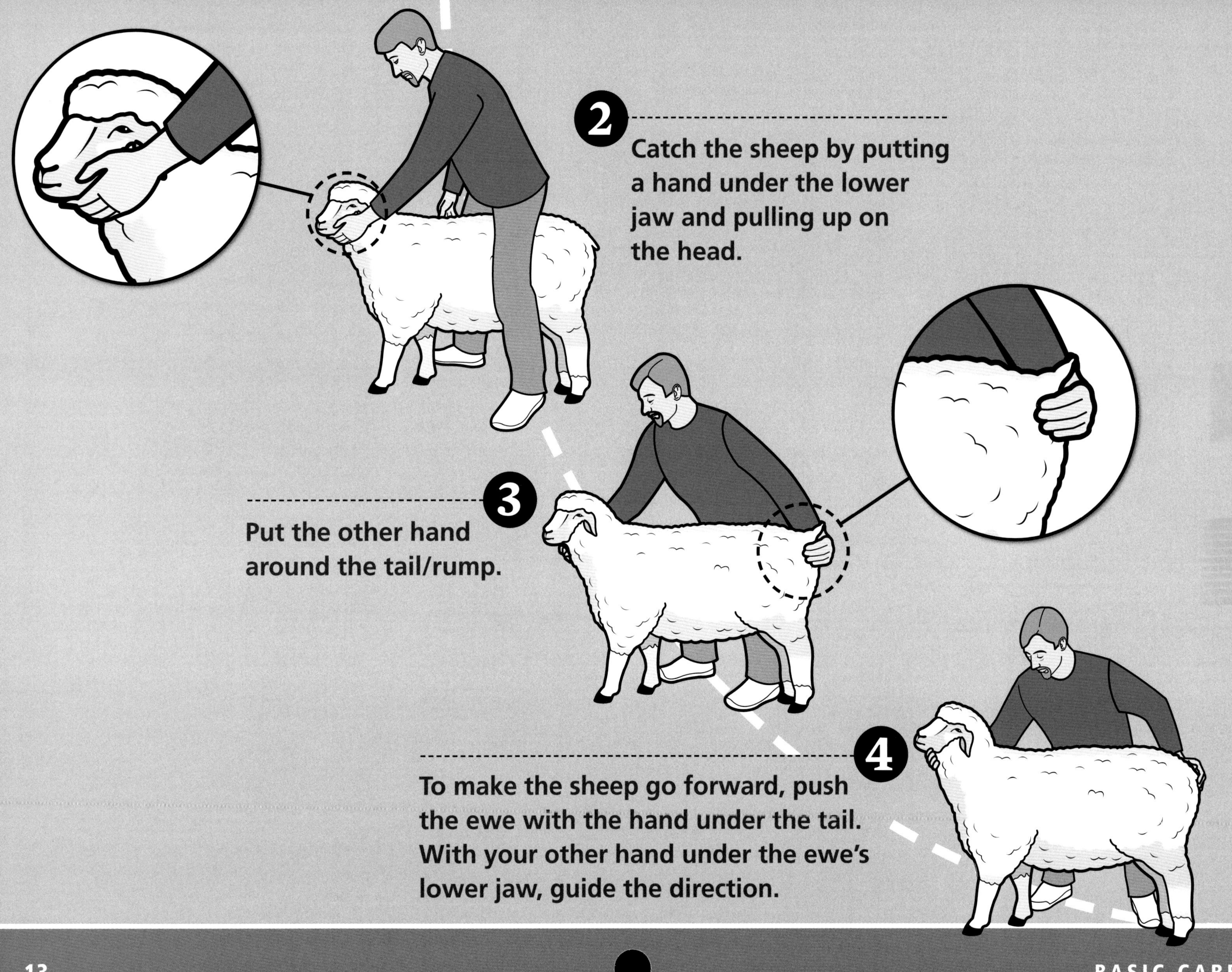
2
Catch the sheep by putting a hand under the lower jaw and pulling up on the head.
3
Put the other hand around the tail/rump.
4
To make the sheep go forward, push the ewe with the hand under the tail. With your other hand under the ewe's lower jaw, guide the direction.

# Throwing a Sheep

**THE BEST WAY,** in general, to perform tasks such as trimming hooves and shearing or to address health issues where the sheep must be held stationary is to "throw" the sheep. It will usually sit quietly when placed on its rump and held in a secure and stable position.

**Slip your left thumb into the sheep's mouth behind the incisor teeth and place your other hand on the sheep's right hip.**

## TIPS FOR HANDLING SHEEP

- Don't move a sheep too far away from the rest of the flock. An isolated sheep will become easily stressed.
- Always work quietly around your sheep; this will make it easier to catch and throw.
- Beware of rams that may butt you when your back is turned.
- Well-trained herding dogs can be very effective at moving sheep.
- Sheep may follow a bucketful of grain.
- For a large flock, consider building a chute. The sheep can walk through it in single file for easy handling.

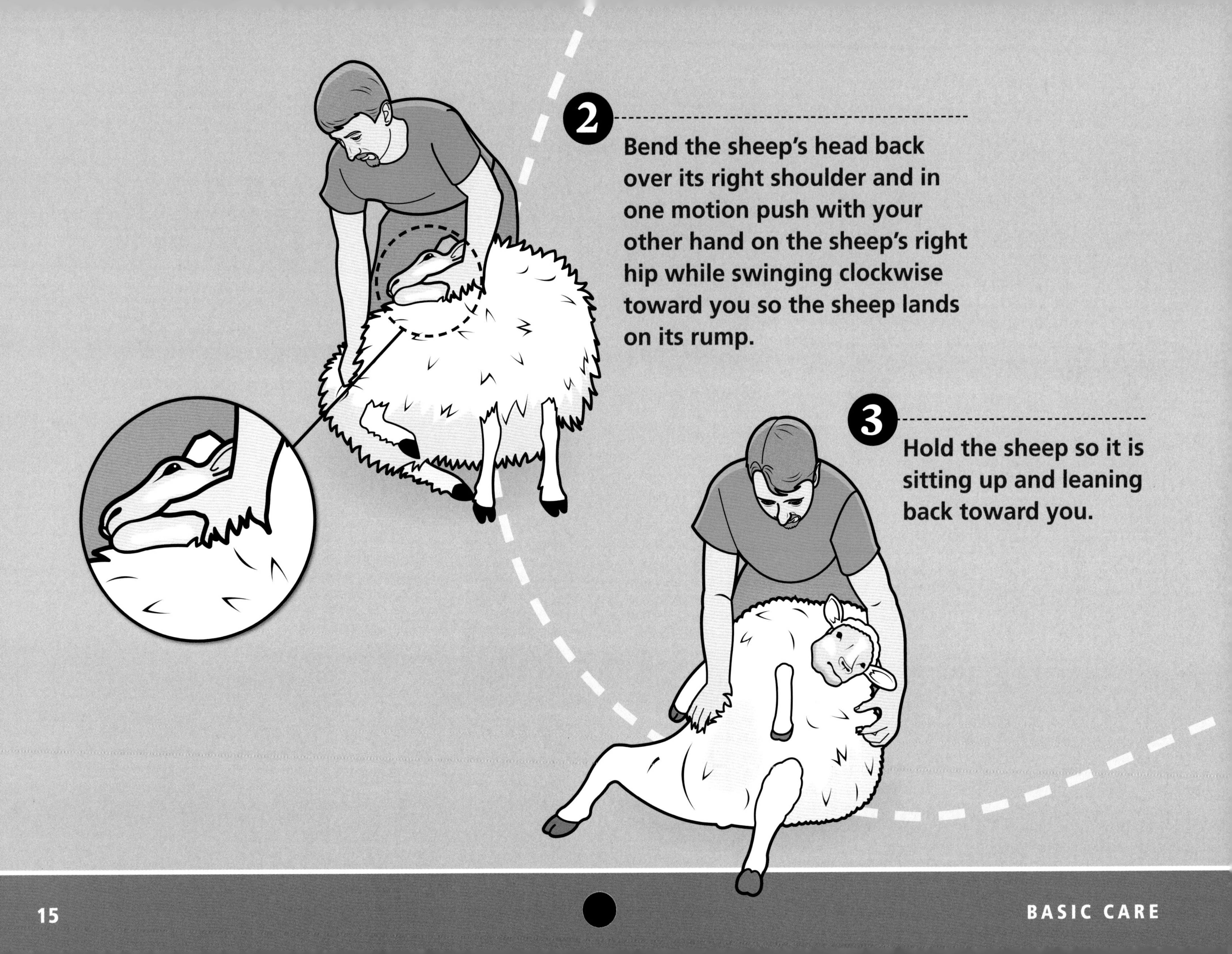

**2** Bend the sheep's head back over its right shoulder and in one motion push with your other hand on the sheep's right hip while swinging clockwise toward you so the sheep lands on its rump.

**3** Hold the sheep so it is sitting up and leaning back toward you.

# Trimming a Hoof

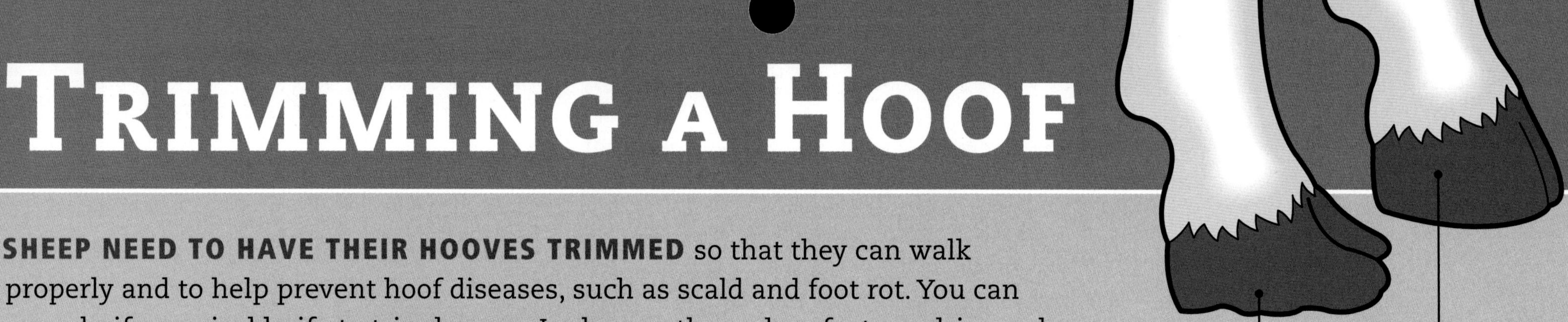


**SHEEP NEED TO HAVE THEIR HOOVES TRIMMED** so that they can walk properly and to help prevent hoof diseases, such as scald and foot rot. You can use a knife or a jackknife to trim hooves. In dry weather, when feet are drier and harder to trim, hoof shears can be useful. How often you trim your sheep's hooves depends on the conditions of your pasture and barn floor. Some farmers need to trim their sheep's hooves only once a year; others may need to do it twice a year.

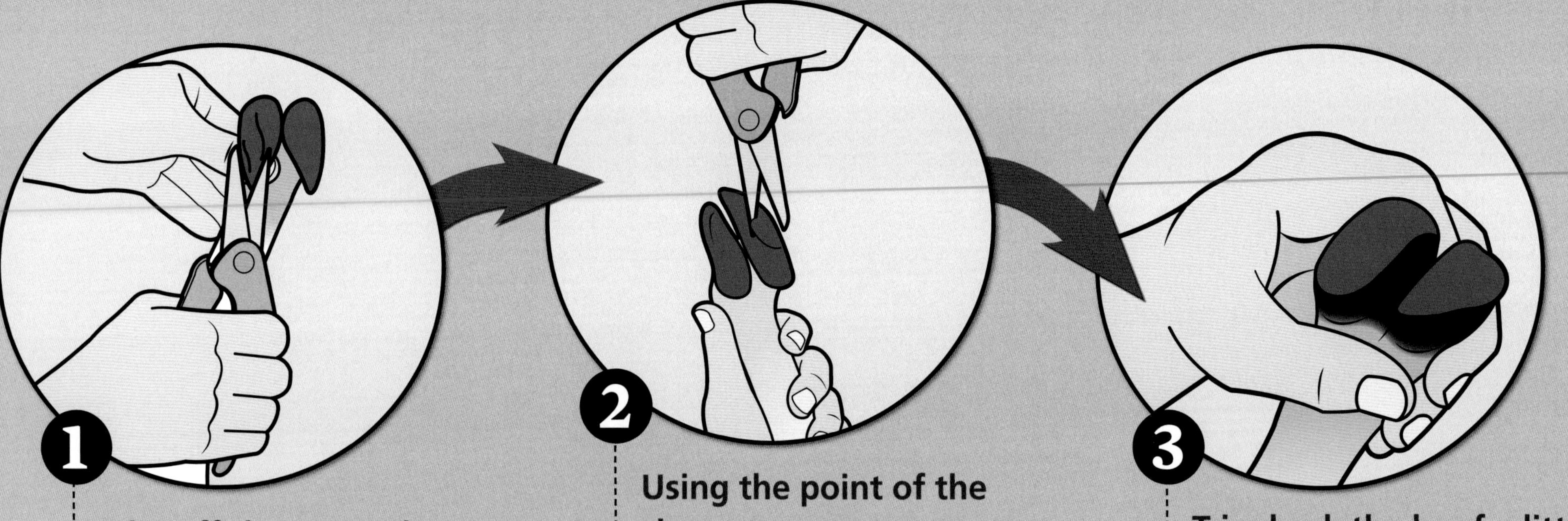


1. Trim off the excess horn so that it is level with the sole and does not protrude too far in front.
2. Using the point of the shears, scrape out any pockets where mud or manure has gathered.
3. Trim back the hoof a little farther to the level of the foot pad. *(Notice the shape of the hooves on half-grown lambs.)*

# TAKING TEMPERATURE

**IT IS IMPORTANT TO RECOGNIZE** the signs that your sheep is not well. Signs of abnormality include loss of appetite, not coming to eat as usual, and standing apart from the group. Any weakness or staggering, unusually labored or fast breathing, change in bowel movements, change in "personality," and a temperature of more than 103˚F (40˚C) all indicate possible problems. Using a veterinary rectal thermometer or a regular thermometer, follow these steps to take your sheep's temperature. Unless the weather is very hot, the sheep's temperature should be between 100.9˚F (38.3˚C) and 103˚F (39.4˚C).

1. **Tie a long string to the ring, hole, or edge of the outer end of the thermometer.**
2. **Holding the sheep, fully insert the thermometer into the rectum.**
3. **Leave the thermometer in for 2 to 3 minutes, then remove.**

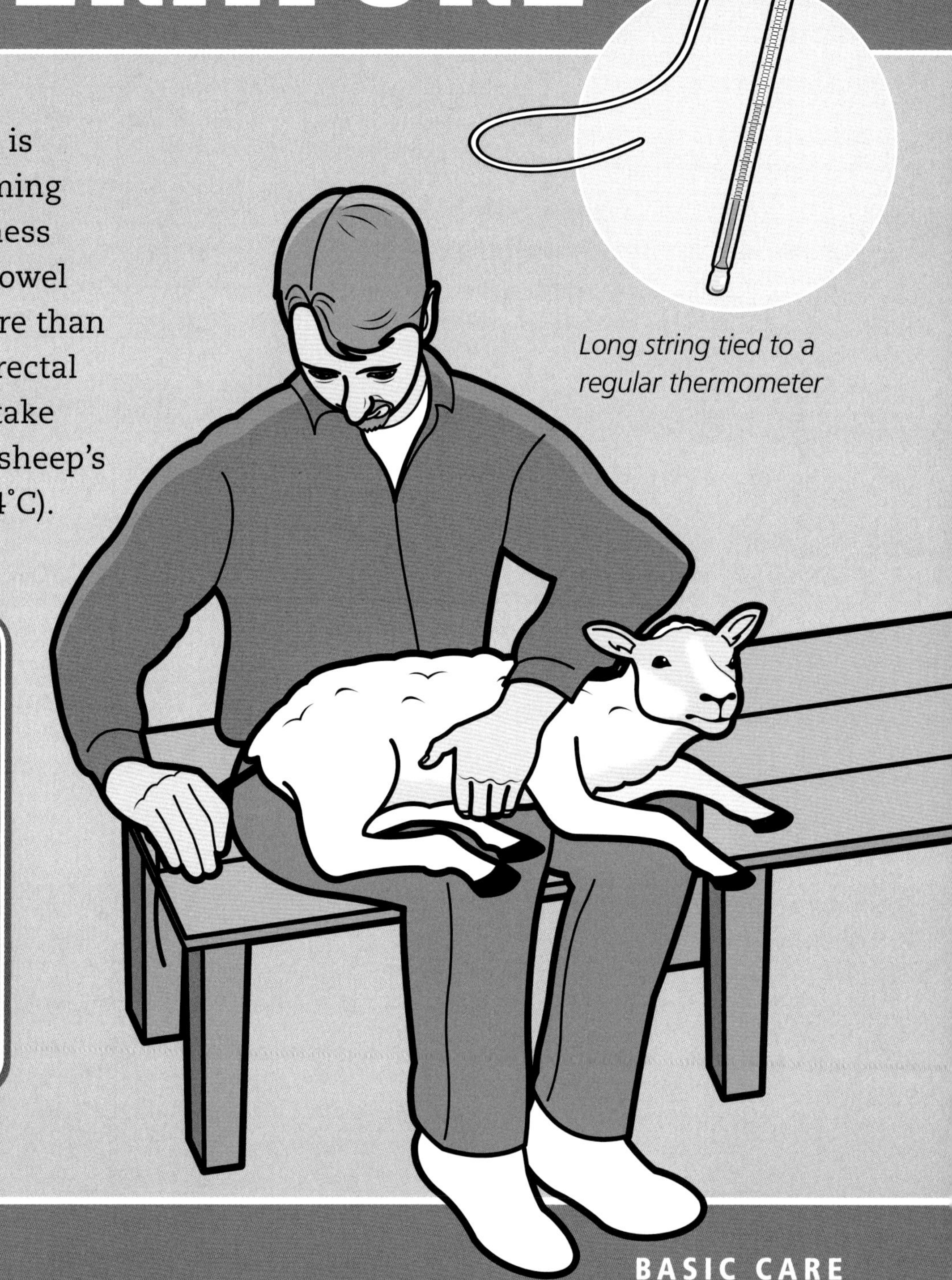

*Long string tied to a regular thermometer*

# Giving Injections

## SUBCUTANEOUS

Most vaccinations are given by this route. The medication should usually be at body temperature, especially with young lambs. It can be given in the neck, but the preferred place is in the loose, hairless skin behind and below the armpit, over the chest wall. Be careful not to inject into the armpit. This can happen if the injection is made too far forward.

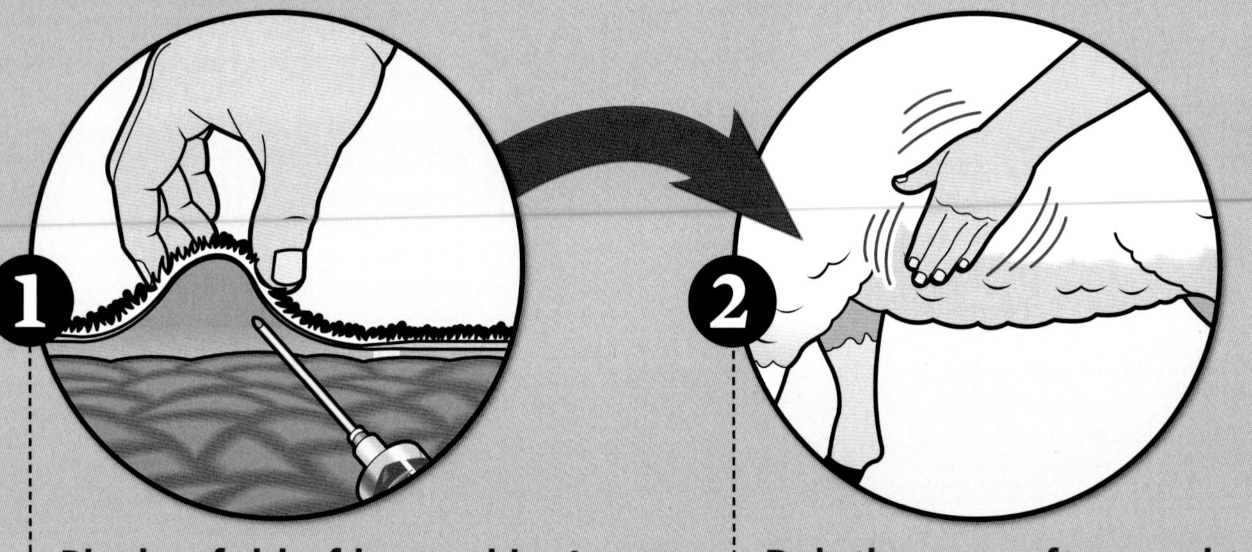


**1. Pinch a fold of loose skin. Insert the needle under the skin, holding it parallel to the body surface. Retract the plunger about ¼ inch (roughly ¾ centimeter), then push it in all the way.**

**2. Rub the area afterward to distribute the medication and hasten absorption.**

### TIPS

- **Distribute doses of more than 10 mL among several sites instead of in one place.**
- **Do not inject near a joint, in areas with more than a small amount of fat under the skin, or into a muscle.**
- **Veins are usually not a problem, but if you want to make sure you are not in a vein, pull out the plunger slightly before injecting. If it draws out blood, try another spot.**

## INTRAMUSCULAR

Many antibiotics, such as penicillin, should be injected into the sheep's muscle. The best sites for administration are along the neck, into the muscle. Be certain you are well over on one side or the other and not toward the vertebrae and spinal cord.

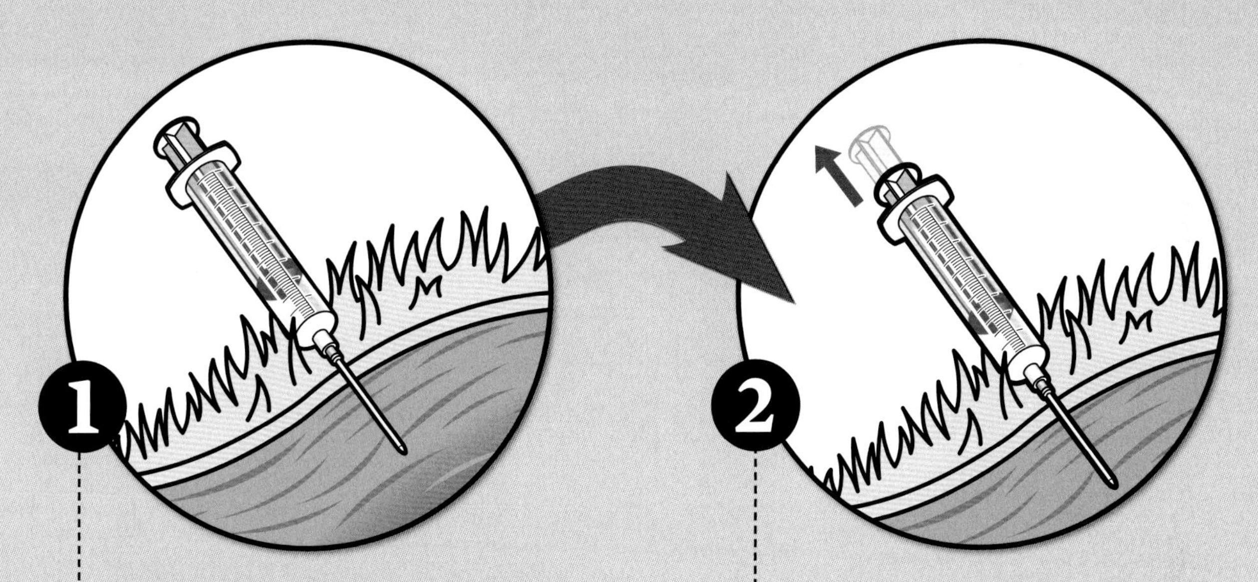


1. Have an assistant hold the sheep still. Thrust the needle quickly into the muscle, where indicated above, left.

2. To make sure you haven't accidentally entered a vein, pull back slightly on the plunger. If blood enters the syringe, the needle is in a vein; withdraw it and reinsert.

### TIPS

- Distribute doses of more than 10 mL among several sites instead of in one place.
- Any intramuscular injection runs the risk of creating a lesion in meat and will have to be cut out. Therefore, it is best to give the injection in the neck area.

# Body Condition

**YOUR SHEEP MUST BE IN GOOD CONDITION** for production (breeding, late pregnancy, lactation). A body-condition score assesses the fat and muscle around the vertebrae in the loin region. Scores range from 1 to 5, with 1 being emaciated and 5 being obese. Follow the steps to the right and then determine you sheep's score using the chart on the page below.

| PRODUCTION STAGE | OPTIMAL SCORE |
| --- | --- |
| Breeding | 3.0–4.0 |
| Early to mid-gestation | 2.5–4.0 |
| Lambing | |
| (singles) | 3.0–3.5 |
| (twins) | 3.5–4.0 |
| Weaning | 2 or higher |

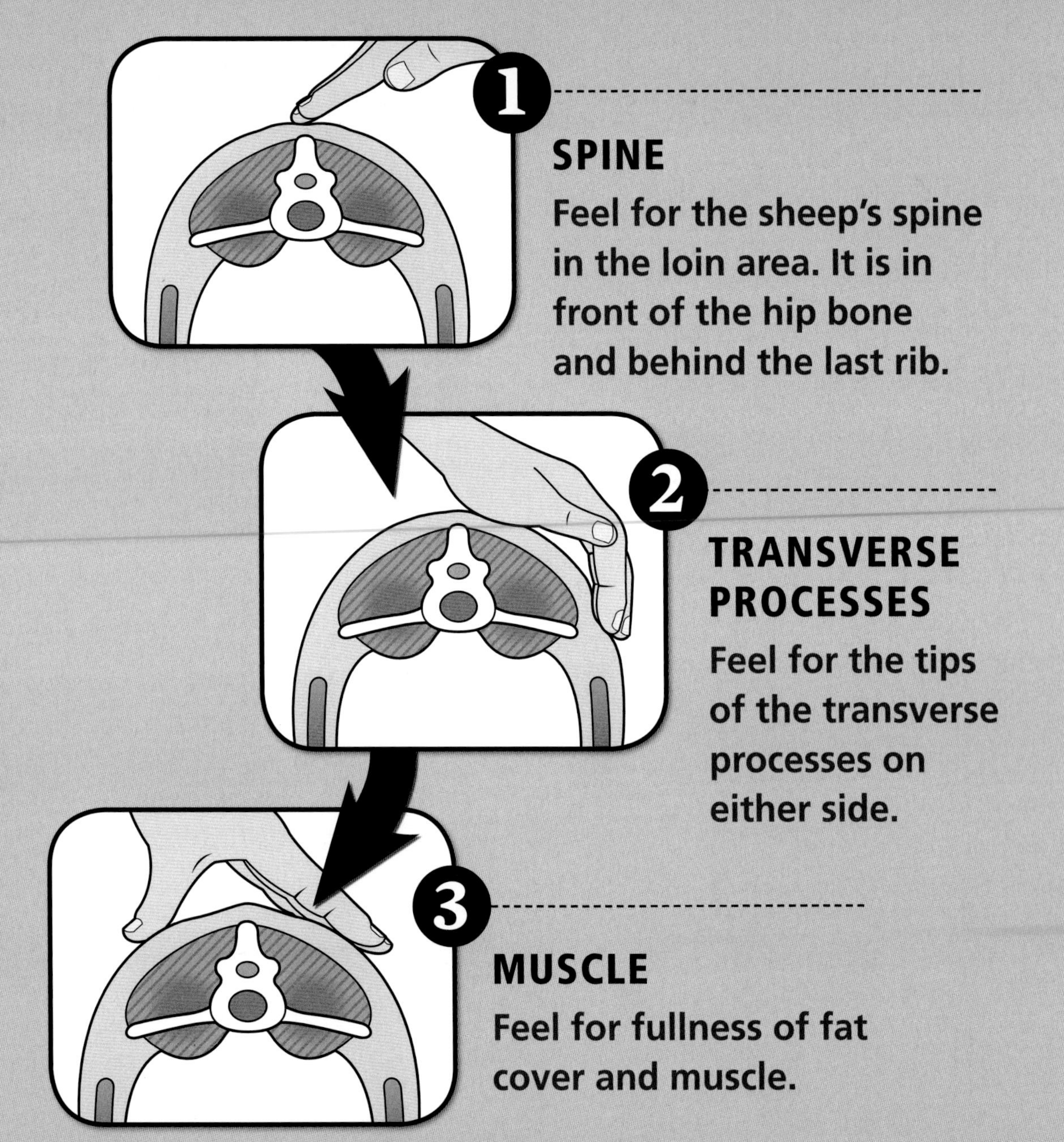

# BODY CONDITION SCORES

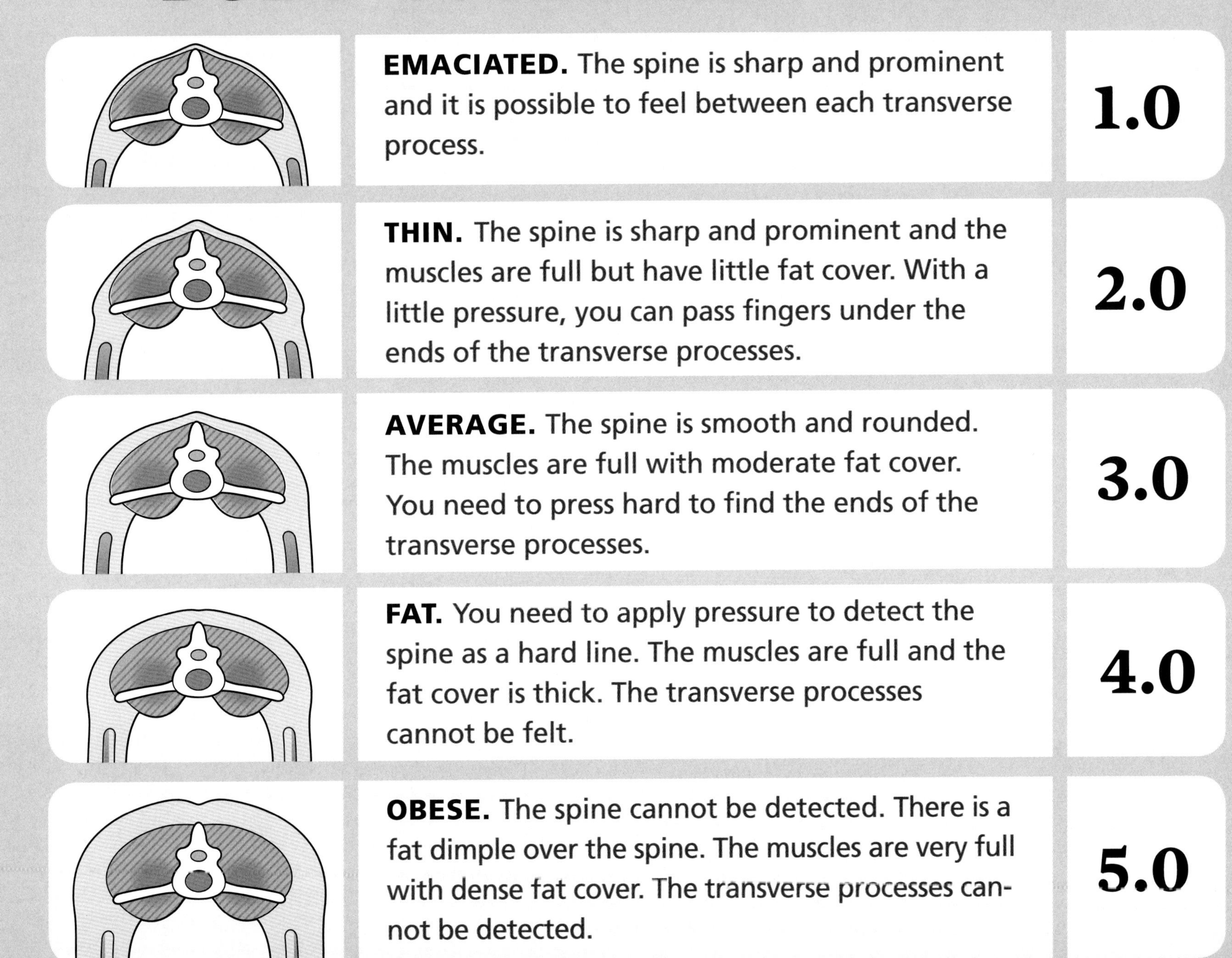

| Description | Score |
|---|---|
| **EMACIATED.** The spine is sharp and prominent and it is possible to feel between each transverse process. | 1.0 |
| **THIN.** The spine is sharp and prominent and the muscles are full but have little fat cover. With a little pressure, you can pass fingers under the ends of the transverse processes. | 2.0 |
| **AVERAGE.** The spine is smooth and rounded. The muscles are full with moderate fat cover. You need to press hard to find the ends of the transverse processes. | 3.0 |
| **FAT.** You need to apply pressure to detect the spine as a hard line. The muscles are full and the fat cover is thick. The transverse processes cannot be felt. | 4.0 |
| **OBESE.** The spine cannot be detected. There is a fat dimple over the spine. The muscles are very full with dense fat cover. The transverse processes cannot be detected. | 5.0 |

# Docking a Tail

**IF YOU'RE LAMBING IN A BARN,** tails of heavy wool breeds should be docked (removed) before the lambs are turned out. Sheep of most breeds are born with long tails, and these can accumulate large amounts of manure in the wool, attracting flies and then maggots (fly-strike). In other words, the tails can serve as a general source of filth, interfering with breeding, lambing, and shearing.

Here is the easiest way to dock a tail. Lambs should be at least 24 hours old but no older than 1 week. Vaccinate against tetanus. A tetanus vaccine given to ewes 2 weeks before lambing protects lambs for about 6 weeks after birth.

**1** **Store the elastrator rubber rings in a small, wide-mouth jar of alcohol, disinfectant, or mild bleach solution. Reach for one, disinfecting your fingers as you do so.**

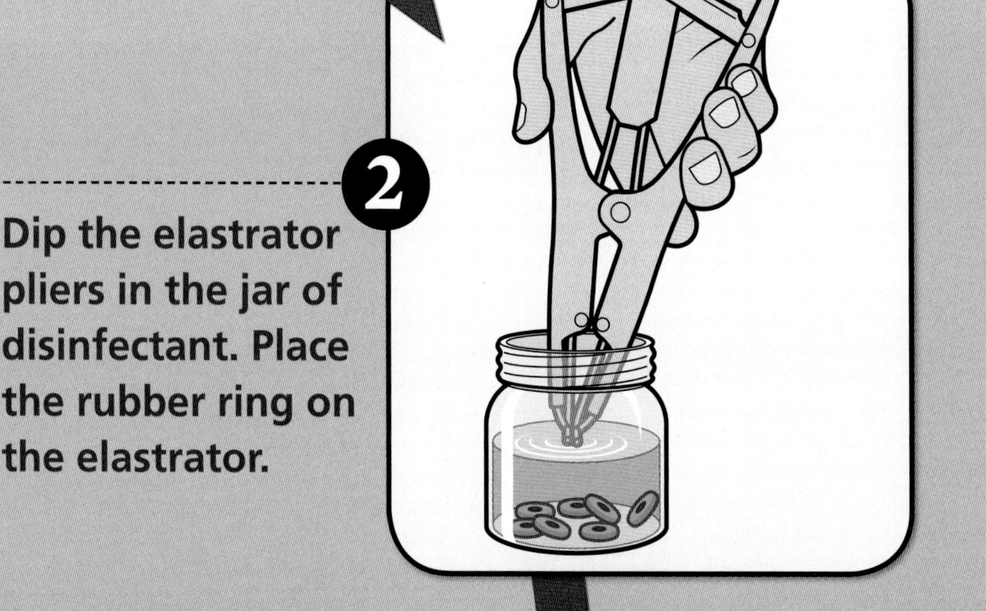

**2** **Dip the elastrator pliers in the jar of disinfectant. Place the rubber ring on the elastrator.**

## OTHER DOCKING METHODS

- Cutting with a dull knife
- Knife and a hammer over a wooden block
- Hot electric chisel or clamp
- Burdizzo emasculator and knife

**3** Place the elastrator around the tail, about 1½ inches (roughly 3.75 centimeters) from the body.

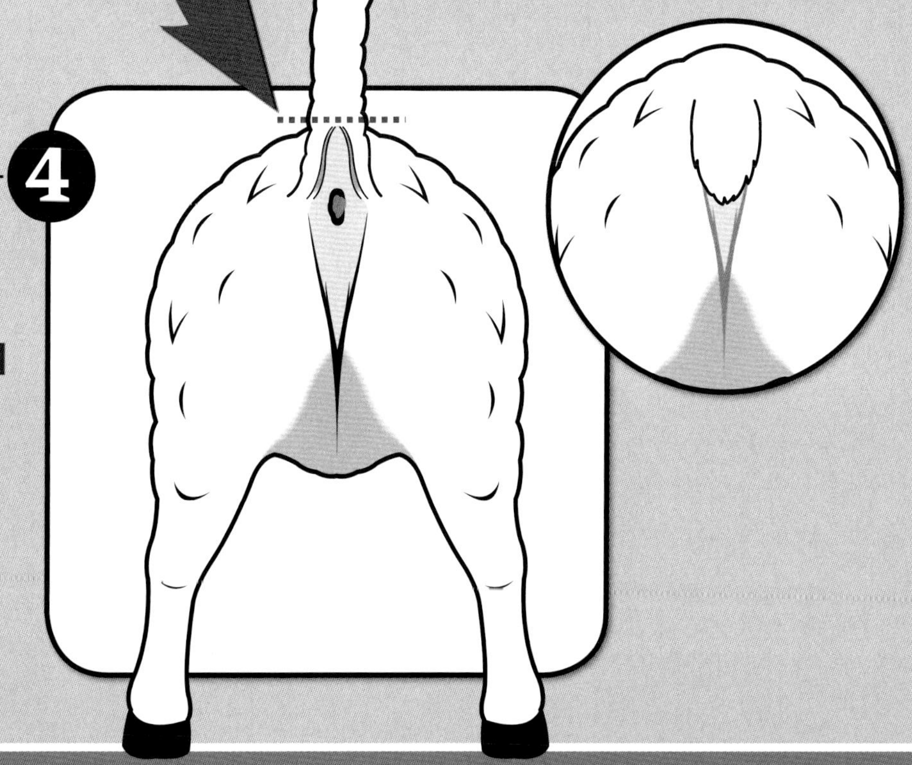


**4** Lift the tail and look at the two flaps of skin that attach from the underside of the tail to the area on each side of the rectum. The band should be placed where the skin attaches to the tail. The tail will fall off in 1 to 3 weeks.

# CASTRATING



**CASTRATION CAN BE DONE EARLY,** as soon as the testicles have descended into the scrotum. Here are two popular techniques for castration. You may wish to have an assistant restrain the animal, so that you can focus on the procedure.

## EMASCULATOR METHOD

This techinique should be used when the lamb is between 1 and 2 weeks of age. The emasculator crushes the sperm cords before it cuts them, thus preventing serious hemorrhage.

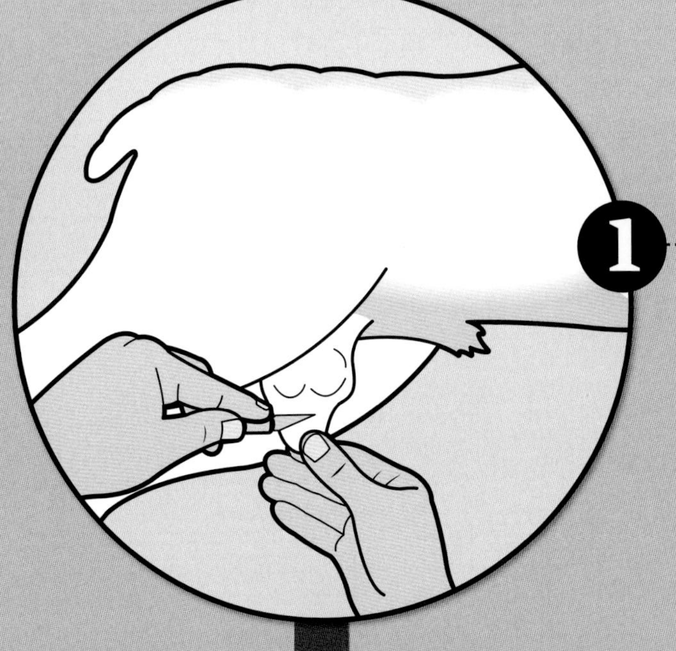

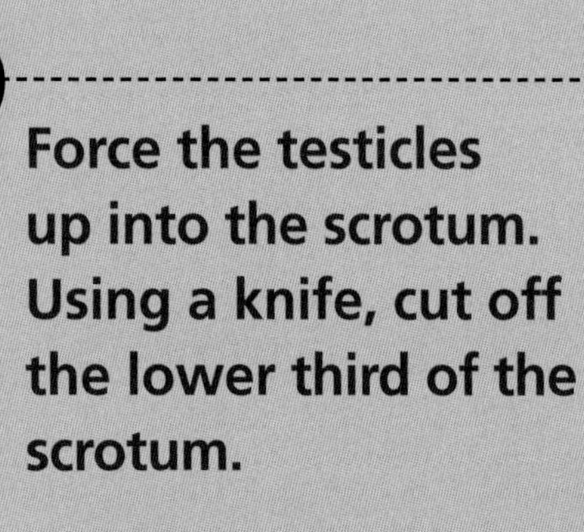

**1** Force the testicles up into the scrotum. Using a knife, cut off the lower third of the scrotum.

### SHOULD YOU CASTRATE?

**YES**

- If you plan to sell to a packinghouse (you will be penalized for not castrating)
- If you intend to keep the rams for longer than 6 months before slaughter
- If rams will be kept with ewes longer than 4 months, especially with breeds that are known to cycle year-round

**NO**

- If you have early lambs and plan on selling the rams for meat at 5 months of age (before breeding season)
- If rams younger than 4 months will be separated from the ewes

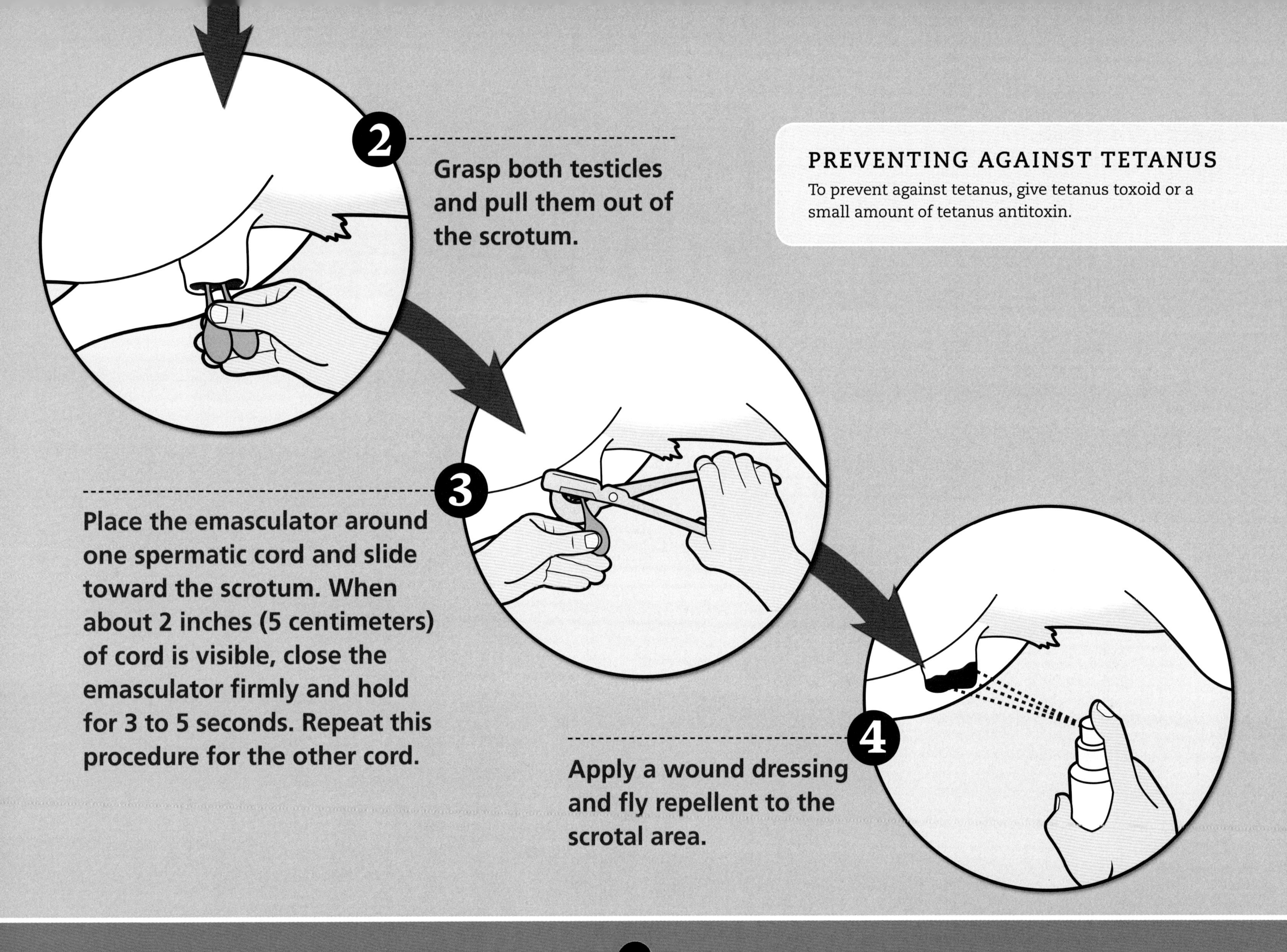


## PREVENTING AGAINST TETANUS

To prevent against tetanus, give tetanus toxoid or a small amount of tetanus antitoxin.

# CASTRATING



**THE ELASTRATOR METHOD** shown below is most often used by first-time farmers. No cutting is involved, so you do not need to dress a wound.

## ELASTRATOR METHOD

This technique can be used for castration when a lamb is about 10 days old. If you have problems with infection, douse the band with iodine after a week. In hot weather, you can spray it with fly repellent.

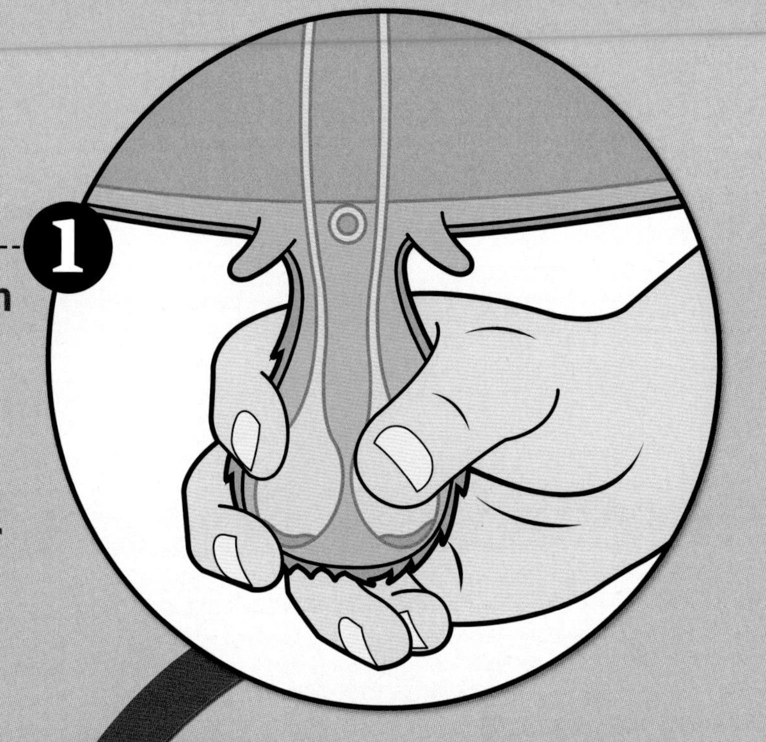

**1** Grasp the scrotum and feel to make sure that both testicles have descended into it.

### CRYPTORCHID
*(OR SHORT SCROTUM)*

The elastrator ring can also be used as a means of sterilization. The rubber elastrator ring is used on the scrotum, but the testes are pushed back up into the body cavity. The increased heat on the sperm results in sterilization, but the animal may still show sexual activity. The male hormones are still present to increase weight gain with more lean meat. This method is used at about 4 weeks of age and the animal is called a cryptorchid (meaning "hidden testicles").

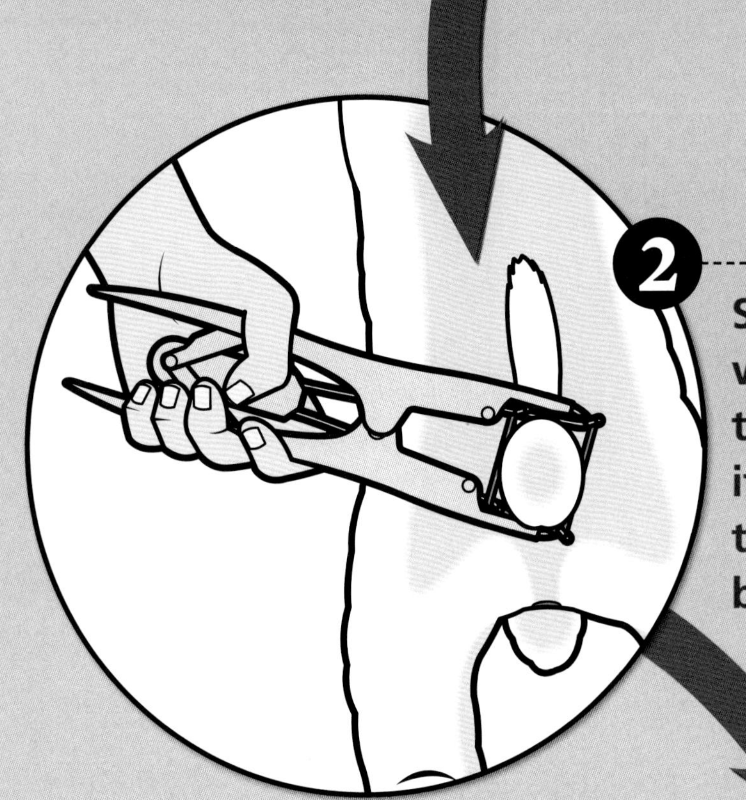


Stretch the rubber ring with the pliers and pull the scrotum through it. Make sure that both testicles are below the band.

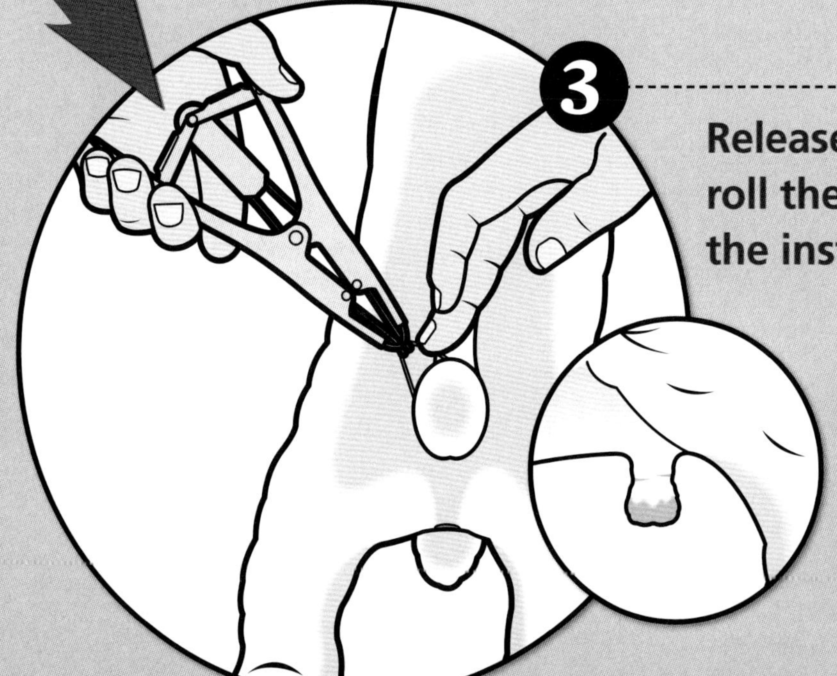


Release the handles and roll the rubber band off the instrument.

The testicles will wither within 30 to 40 days.

### WHY CASTRATE?

The purpose of castration is not only to prevent reproduction. It also improves the quality of the meat and makes the animal calmer and easier to handle.

CHAPTER THREE

# Feeding

# 10 TOXIC SUBSTANCES

**MANY TYPES OF SUBSTANCES CAN BE HARMFUL TO LIVESTOCK.** Store all chemicals and cleaning supplies where animals can't get into them, and always properly dispose of waste containers! Be sure to store or dispose of the following substances.

1. **OLD CRANKCASE OIL & WASTE MOTOR OIL** (high lead content)

2. **OLD RADIATOR COOLANT OR ANTI-FREEZE** (sweet and attractive to sheep)

3. **WEED SPRAY** (some have a salty taste) **& ORCHARD SPRAY DRIPPED ONTO GRASS**

4. **MOST SHEEP INSECTICIDAL DIPS & SPRAYS**

5. **OLD PESTICIDE & HERBICIDE CONTAINERS FILLED WITH RAINWATER**

6. **OLD CAR BATTERIES** (sheep like the salty taste of lead oxide)

7. **SALT** (ingesting too much will cause salt poisoning)

8. **COMMERCIAL FERTILIZER.** Be sure not to spill fertilizer where sheep can eat it, and store the bags carefully. Sheep may nibble on an empty bag. Several rainfalls are needed after fertilizing a field, and it still may not be safe unless the pasture grass is supplemented with grain and hay. Signs of fertilizer toxicity are weakness, rapid open-mouthed breathing, and convulsions.

9. **COW SUPPLEMENTS CONTAINING COPPER** (contain levels of copper lethal for sheep)

10. **"EMPTY" LEAD BUCKETS FILLED WITH RAINWATER** (rainwater is soft and readily dissolves lead; lead-laced rainwater can kill a sheep that drinks it)

# DIGESTIVE SYSTEM

**DIGESTION IN SHEEP IS A COMPLEX PROCESS** that takes place in a four-stomach system. The stomachs are the rumen, reticulum, omasum, and abomasum.

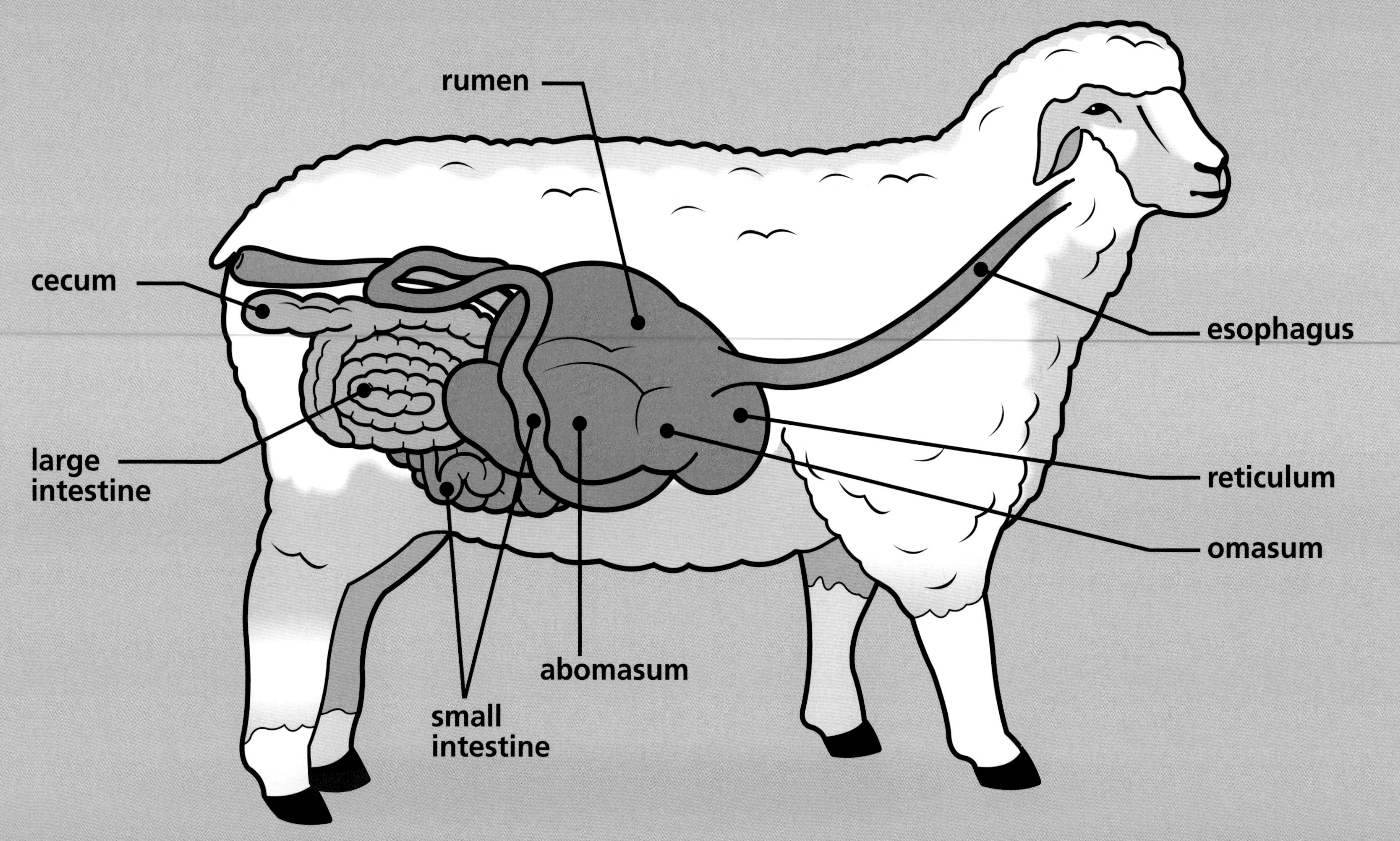

# Feed Requirements

## For Growing Sheep

| Live weight (lbs) | 50.00 | 75.00 | 100.00 | 125.00 |
|---|---|---|---|---|
| Dry matter (lbs) | 2.20 | 3.50 | 4.00 | 4.60 |
| Crude protein (%) | 12.00 | 11.00 | 9.50 | 8.00 |
| Crude protein (lb) | 0.26 | 0.39 | 0.38 | 0.37 |
| TDN (%) | 55.00 | 58.00 | 62.00 | 62.00 |
| TDN (lbs) | 1.21 | 2.03 | 2.48 | 2.85 |
| Energy (Mcal) | 1.14 | 1.18 | 1.27 | 1.27 |
| Calcium (%) | 0.23 | 0.21 | 0.19 | 0.18 |
| Phosphorus (%) | 0.21 | 0.18 | 0.18 | 0.16 |

**TDN** = total daily nutrients

***Note:*** *Don't feed sheep feed formulas or mineral mixtures that are not specifically recommended for them. The amounts of some trace minerals, such as copper, that are in feed for other classes of livestock are toxic to sheep.*

## For Breeding Sheep

| | First Two-Thirds of Gestation | Last One-Third of Gestation | First 10 Weeks of Lactation | Last 14 Weeks of Lactation | Rams at Moderate Work |
|---|---|---|---|---|---|
| Dry matter (lb per 100 lb of body weight) | 2.50 | 3.50 | 4.20 | 3.50 | 3.50 |
| Crude protein (%) | 8.00 | 8.20 | 8.40 | 8.20 | 7.60 |
| Crude protein (lb per 100 lb of body weight) | 0.20 | 0.29 | 0.35 | 0.29 | 0.27 |
| TDN (%) | 50.00 | 52.00 | 58.00 | 52.00 | 55.00 |
| TDN (lb per 100 lb of body weight) | 1.25 | 1.82 | 2.44 | 1.82 | 1.93 |
| Energy (Mcal per lb of feed) | 1.00 | 1.10 | 1.20 | 1.10 | 1.20 |
| Calcium (%) | 0.24 | 0.23 | 0.28 | 0.25 | 0.18 |
| Phosphorus (%) | 0.19 | 0.17 | 0.21 | 0.19 | 0.16 |

# Toxic Plants

**SHEEP ARE CURIOUS AND LOVE TO NIBBLE ON STRANGE PLANTS,** especially when pasture is lacking other, necessary forage, so you must prevent them from contact with any poisonous plants in their grazing area. Talk to your county Extension agent and investigate potentially dangerous ornamental plantings. Here are the most common plants that are poisonous to sheep.

## OAK

Oak toxicity occurs in the fall, if sheep eat the acorns, or in the spring, if they eat the buds. Oak toxicity affects the kidneys and occurs only if oak is their only diet.

## AMERICAN OR JAPANESE YEW

The needles of this plant are extremely toxic; it should be eradicated from your pasture.

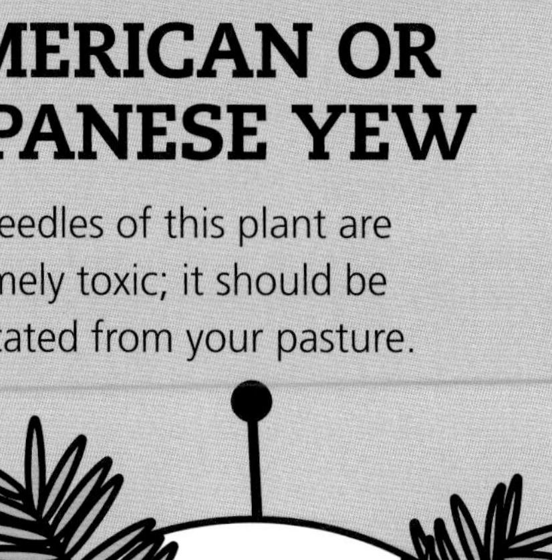

## TANSY RAGWORT

Sheep seldom eat the mature plant, but poisoning can occur when seedlings are grazed or when the plant is mixed with hay.

## RHODODENDRON

All parts of this plant, but particularly the leaves, are very poisonous. Toxicity most often occurs in the winter and early spring, when other foliage is unavailable.

## MILKWEED

This plant is especially poisonous in the summertime, when its seedpods are full.

## NIGHTSHADE

Sheep usually avoid nightshade but will consume it in the fall when grass begins to brown.

## LUPINE

Lupine is especially poisonous in summer and fall.

## SHEEP LAUREL

Similar to rhododendron, sheep laurel is very toxic in all its parts. Keep sheep away from this plant at all times.

## PLAN OF ACTION

1. Walk around the pasture and look for the plants shown above. Note any unusual or unfamiliar plants as well.
2. Send fresh, whole plants you can't identify to the state agricultural college. Wrap the plants in several layers of newspaper before sending.
3. Take the appropriate action, once you know the level of toxicity of your plants. Some plants may be toxic only during certain stages and you need just to keep an eye on them; others may be highly toxic and must be eradicated.
4. Call your veterinarian if you suspect plant poisoning. Keep the sick animal sheltered from heat and cold and allow it to eat only its normal, safe feed.

# PASTURE ROTATION

**BECAUSE GRAZEABLE FORAGE** is the primary source of feed in a pastoral system, learning to keep it growing well is critical. Seeds provide the nutrients required when plants first sprout, but rather quickly the roots must develop and begin supplying all the nutrients for continued growth.

- As the season progresses, the nutrients are used for the development of a flower and seed head.
- As the seed head develops, the root system becomes depleted of its energy supply, and the plant will be vulnerable to additional stress.
- If plants are clipped — either by mechanical means or by an animal biting them — before they begin developing the seed head, the roots keep supplying energy for leaf growth instead of seed production.
- Clipping too much or too often significantly reduces the energy in the root system, thereby slowing leaf growth. The diagram at right, below, shows you where you would ideally clip grazeable forage.
- The optimal frequency of clipping or grazing depends on the time of year, temperature, precipitation, soil fertility, and other factors.

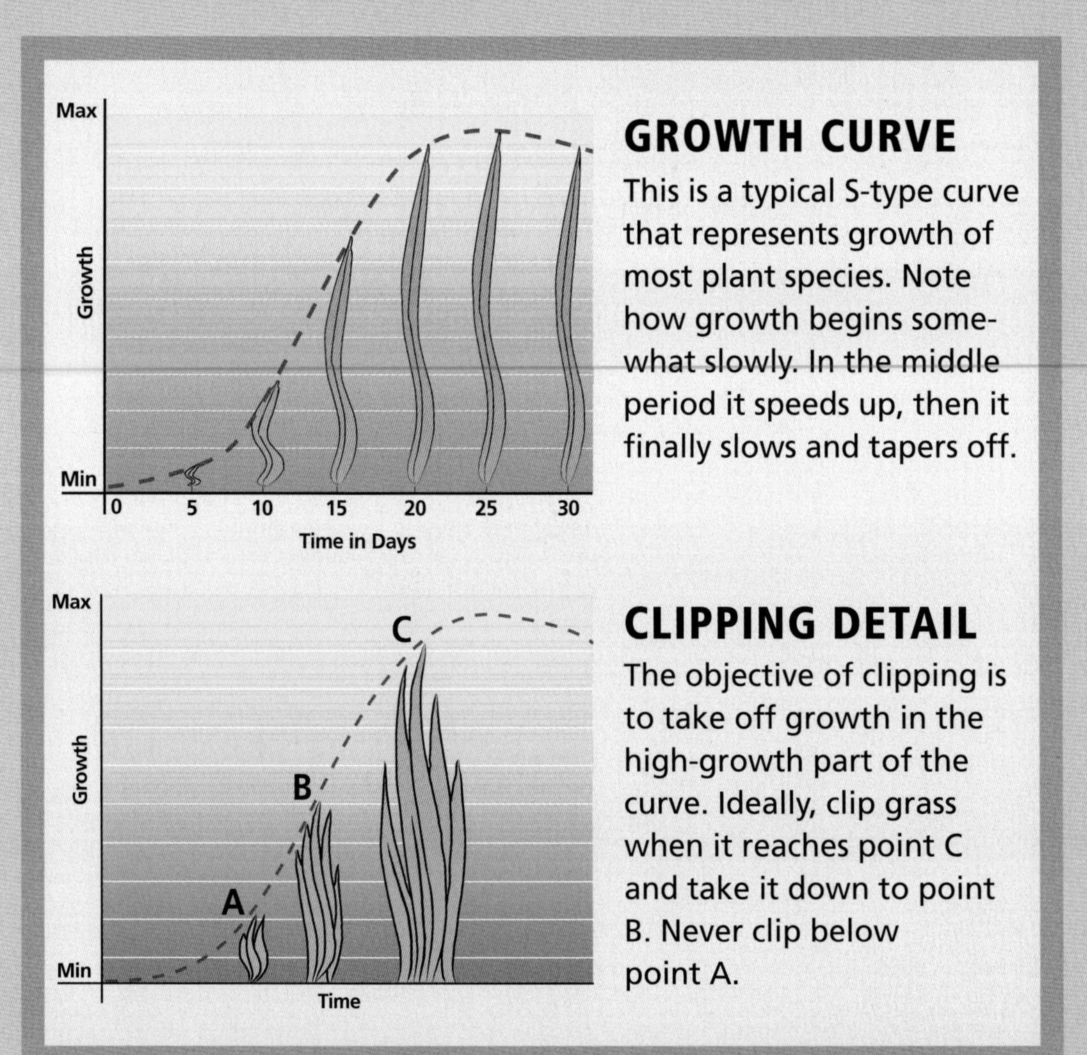


## GROWTH CURVE

This is a typical S-type curve that represents growth of most plant species. Note how growth begins somewhat slowly. In the middle period it speeds up, then it finally slows and tapers off.

## CLIPPING DETAIL

The objective of clipping is to take off growth in the high-growth part of the curve. Ideally, clip grass when it reaches point C and take it down to point B. Never clip below point A.

# PRINCIPLES OF PASTURE ROTATION

**SUBDIVIDE YOUR PASTURE** into smaller pieces, known as paddocks; the more, the better.

**BARRIERS BETWEEN PADDOCKS** can be created with temporary electric netting, polywire, or step-in posts. Temporary fences are excellent for adjusting paddocks for seasonal variations but require more management than do quality permanent fences.

**ANIMALS SHOULD BE MOVED FROM A PADDOCK** before they've grazed off 50 to 60 percent of the forage.

**CONSIDER CLIPPING OR HAYING** if grass grows faster than grazing animals can keep up with. As grasses mature — as early as 8 inches (20 centimeters) tall with some species — they have lower nutritional value and are more easily trampled and wasted.

**THE GRASS SHOULD BE 5 TO 8 INCHES TALL** when the sheep enter the paddock and about 2 inches (5 centimeters) tall when they leave.

**HOW QUICKLY YOU ROTATE YOUR PASTURES WILL VARY ACCORDING TO SEASON.** During the early spring, you can move the flock through the paddocks quickly, leaving several inches of growth before moving on.

**IN THE LATTER PART OF THE GROWING SEASON,** slow down their movements between paddocks so they graze 60 percent of the forage.

**ALLOW THE PADDOCK TO REST.** The time varies according to the season; in the early spring, it may be as short as 7 to 10 days; in the summer, it may take 45 days.

CHAPTER FOUR

# Lambing

# REPRODUCTIVE FUNCTIONS

**HERE ARE SOME RULES OF THUMB** for reproductive functions — but remember, all animals are unique individuals, and some don't follow the rules!

1 **FIRST ESTRUS.** This generally occurs when ewe lambs are at least 6 months of age and weigh two-thirds of their adult weight.

2 **LENGTH OF ESTROUS CYCLE.** The range is 14 to 19 days between cycles; 17 days is the average.

3 **LENGTH OF TIME STANDING IN HEAT.** The average is 30 hours, but this can range anywhere from 3 to 73 hours.

4 **FACTORS AFFECTING MULTIPLE BIRTHS.** Ewes are most likely to give birth to twins if they are genetically disposed; if they receive very good nutrition at the time of breeding; if they are under minimal stress; and if the ram has not been overused. Ewes in their prime (between 2 and 6 years old) are also more likely to have multiple births.

5 **TIME OF OVULATION.** This happens 28 hours after the start of the estrous cycle.

6 **LENGTH OF TIME THE EGG REMAINS VIABLE.** This is between 12 and 24 hours.

7 **LENGTH OF GESTATION.** This is the time between breeding and lambing. In sheep, this period is 147 to 153 days.

8 **EWE'S CYCLE.** Ewes of most breeds of sheep respond strongly to seasonal shifts and will ovulate in response to more hours of darkness. For these sheep, the natural breeding season lasts from late August to February. Some breeds are not affected by seasonal changes.

## Male Reproductive Tract

### The Ram's Role

Use a young ram sparingly for breeding. One way to conserve his energy is to separate him from the ewes for several hours, during which he can be fed and watered and allowed to rest.

One good ram can handle 25 to 30 ewes. In a small flock where the ram gets good feed, you can expect about 6 years of use from him, though you don't want him breeding his daughters and granddaughters indiscriminately. On open range, you may need to replace him after only a couple of years.

For a really small flock, it may not make sense to purchase a ram. You may be able to borrow a neighbor's ram for a small fee.

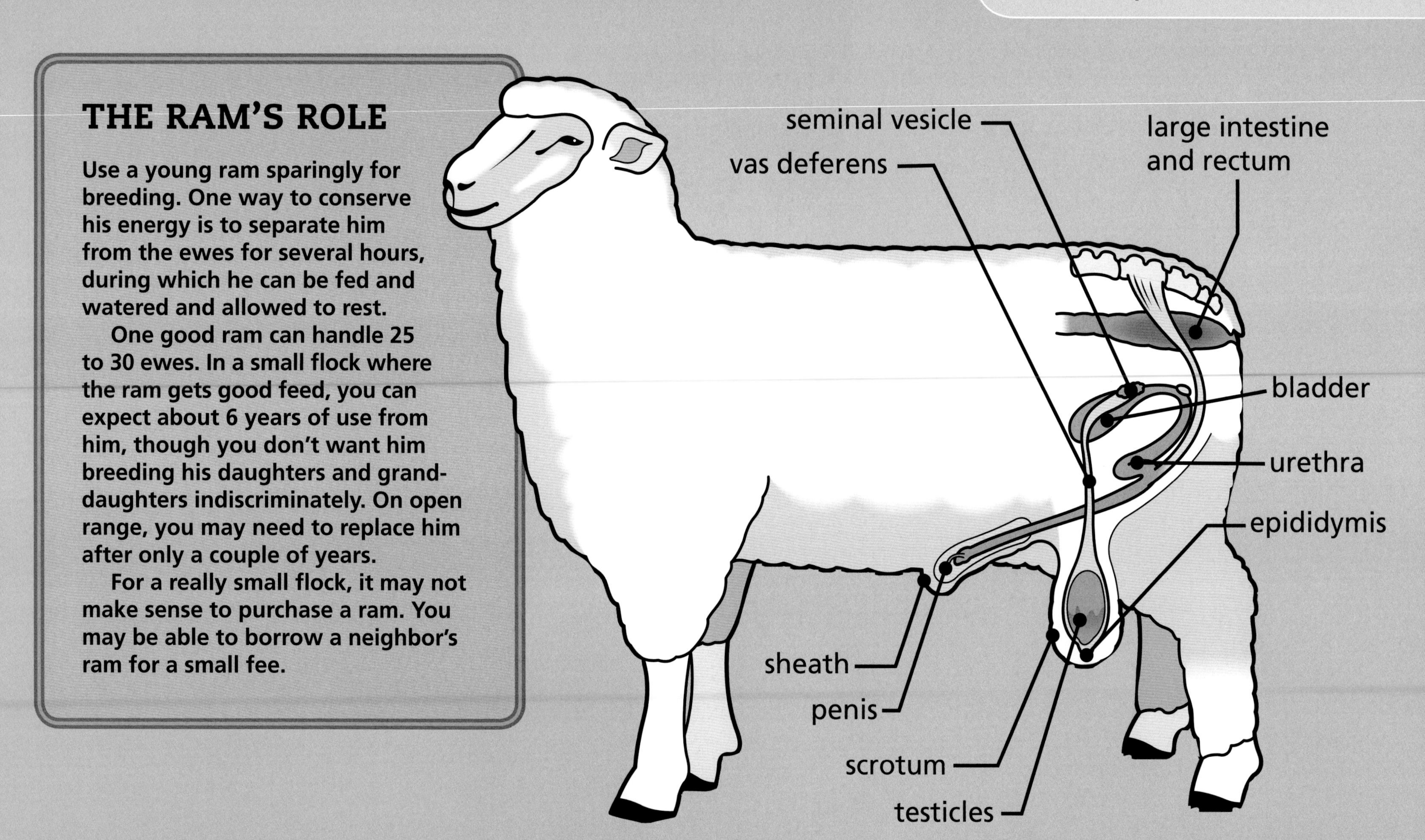

## Female Reproductive Tract

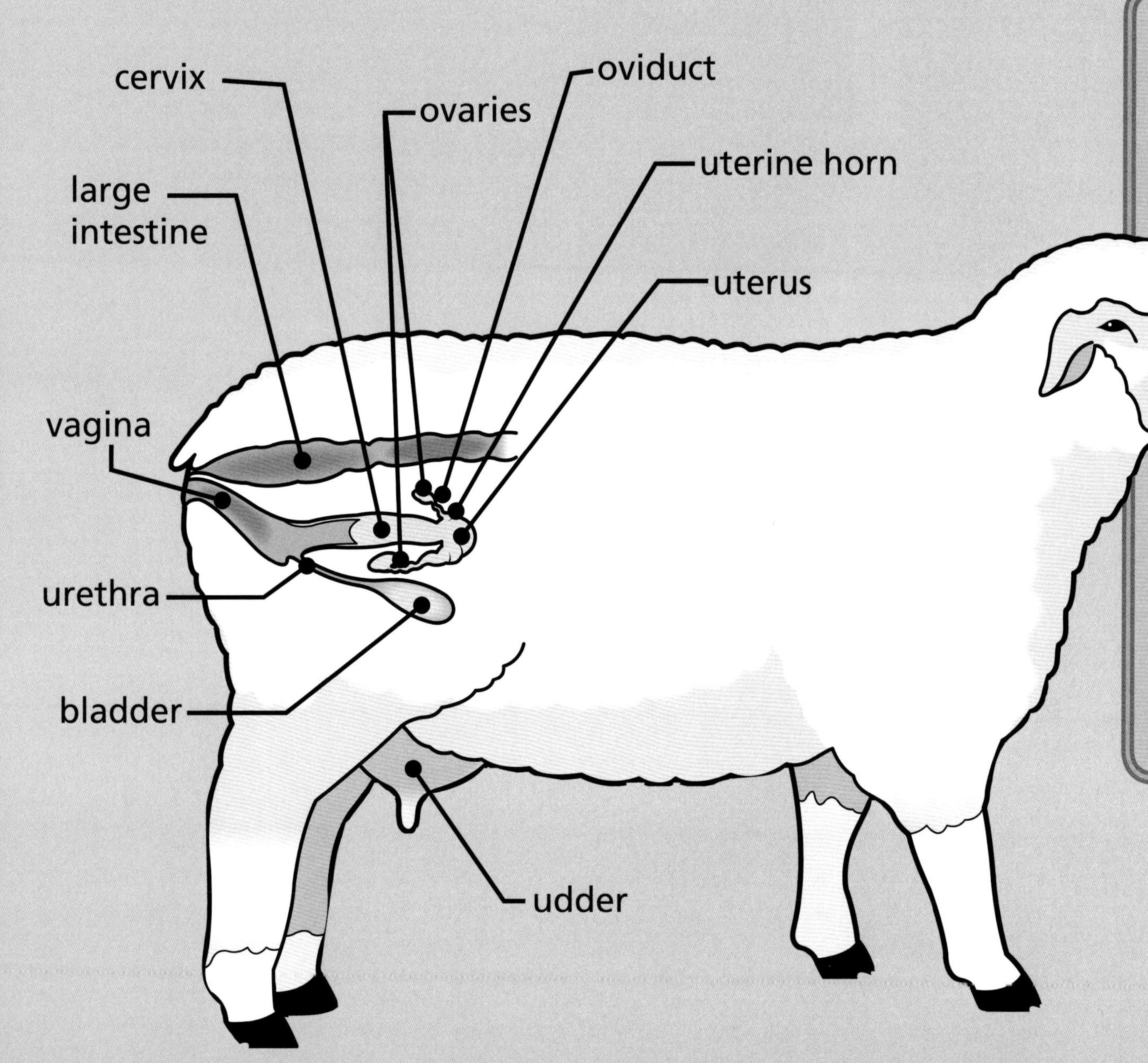


## PREPARING THE EWES

**Before the breeding season begins, trim away any wool tags from around each ewe's tail and trim her feet, because she will be carrying extra weight during pregnancy and it is important for her feet to be in good condition.**

**Seventeen days before you would like to start breeding, put your ram in a pasture adjacent to the ewes with a solid fence separating them. Research has shown that the sound and scent of the ram bring the ewes into heat earlier. Don't pen the ram next to the ewes before this sensitizing period. It is the sudden contact with the rams that excites the females.**

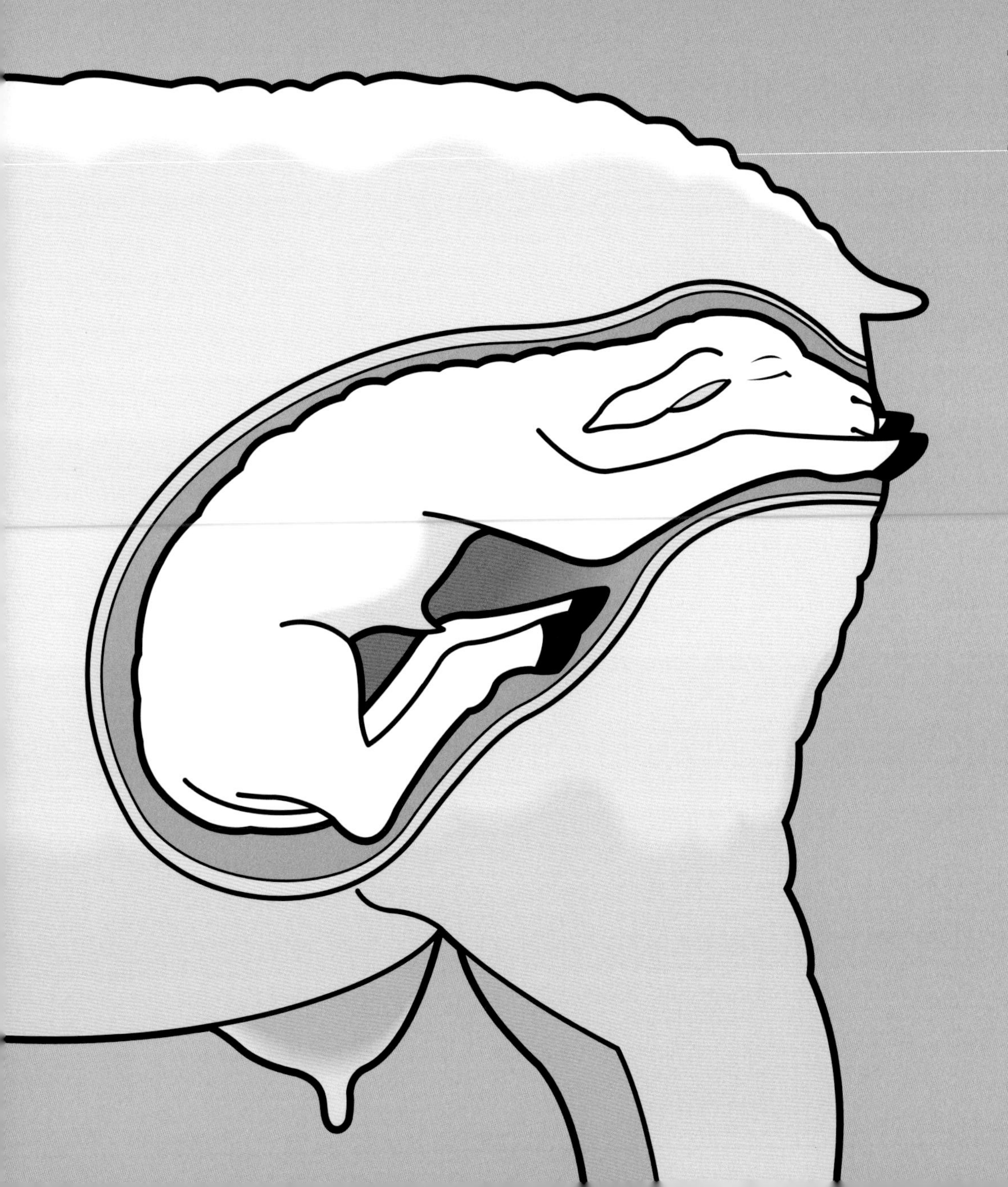

## NORMAL BIRTH

The overwhelming majority of lambs come in the normal, front-feet-first position and require no help. Signs of a normal presentation are:

- **Lamb's nose and front feet are presented**
- **Lamb's back is toward ewe's back**
- **Lamb begins emerging ½–1 hour after the amniotic sac filled with fluid (water bag) is passed out of the vulva**
- **Ewe has steady, strong contractions, usually in a lying-down position**

# LAMBING CHECKLIST

You should always have on hand the following items:

## MEDICAL SUPPLIES

- ☐ 60 mL syringes
- ☐ antibiotics
- ☐ Betadine, to sterilize equipment
- ☐ iodine
- ☐ lubricant, for pulling lambs
- ☐ rubber catheter (tube)
- ☐ rubber gloves
- ☐ tube feeder

## OTHER

- ☐ birth records, breeding records, and due date
- ☐ frozen colostrum (kept in freezer)

## TOOLS & EQUIPMENT

- ☐ 2 small ropes, cable or twine
- ☐ 8-oz (250-mL) or smaller baby bottle
- ☐ bucket and soap
- ☐ docking, tagging, and castration tools
- ☐ dry towels
- ☐ heat lamp
- ☐ plastic sleeves
- ☐ lamb puller
- ☐ scissors (see page 46)

## COLOSTRUM

Colostrum, the first milk produced by a ewe after giving birth, contains antibodies from her immune system. It is essential that lambs receive colostrum as soon as possible after birth, because a lamb's intestinal lining begins shutting down from the moment of birth until it can no longer allow the passage of antibodies. The closing process takes 16 to 48 hours.

If it's not possible for the lamb to receive colostrum from its mother, it must receive it from another ewe. Cow or goat colostrum may be used in a pinch. Commercial preparations are useful when no other colostrum is available. These milk-whey-antibody products transfer some immunity when mixed with a milk replacer.

# Assisting with Lambing

**YOU SHOULD ASSIST** if 2 hours have elapsed since the ewe has passed her water bag and there is little change, or if there is an obviously abnormal presentation. (See pages 44–45.)

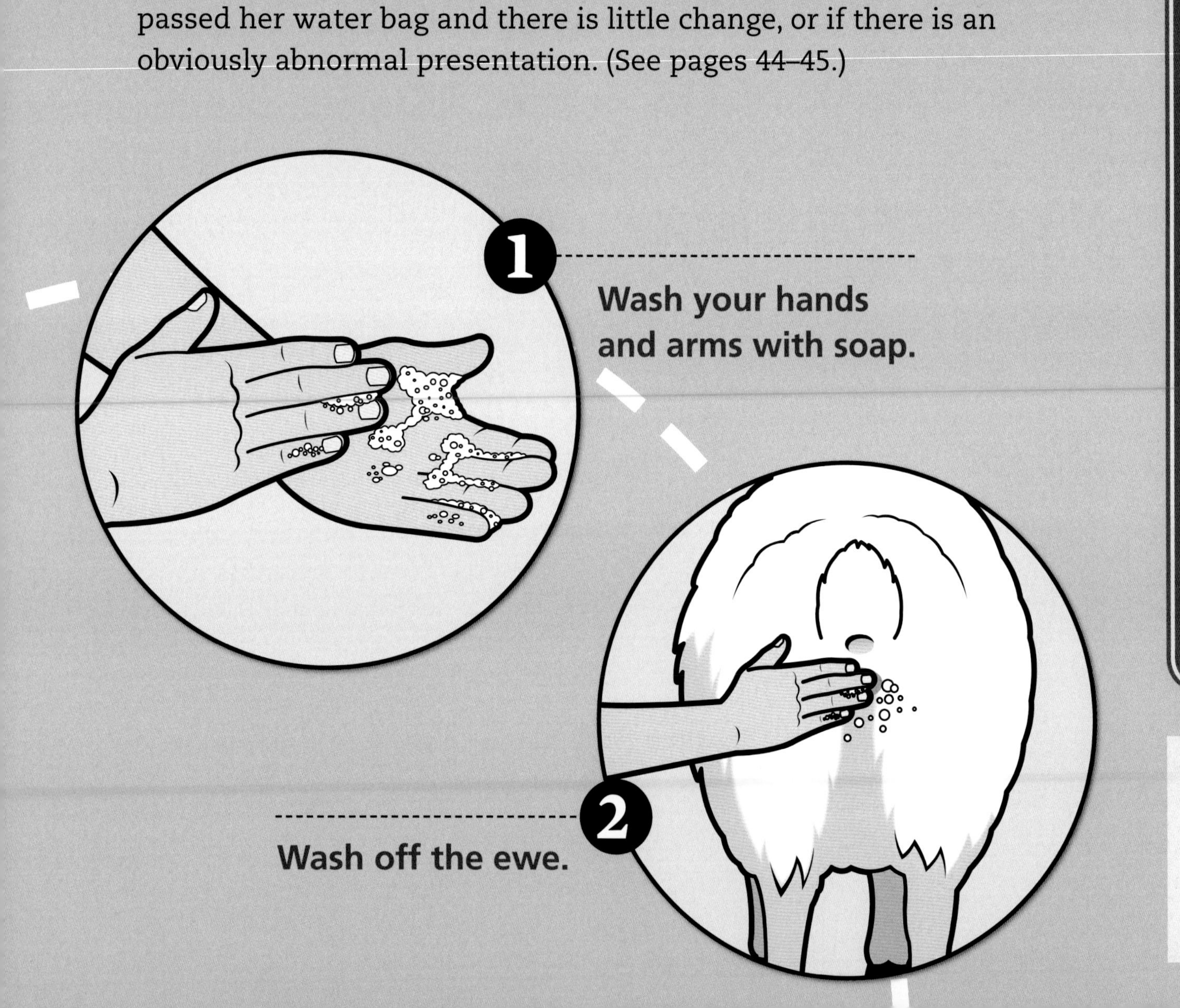


## When to Call the Vet

- If the ewe is obviously in distress
- If the ewe has labored for more than an hour and has made no progress
- If you cannot get the lamb into proper position for delivery
- If a lamb is dead in a ewe and so large it can't be pulled out
- If you feel uncomfortable pulling a lamb

**NOTE:**

If you need to call a veterinarian, bring a pad of paper to take notes and write down questions you may want to ask later.

## WHEN TO ASSIST

As a general rule, let the ewe go on her own unless:

- The lamb's one front leg and nose are both showing but the other front leg is nowhere in sight.
- There are two right or two left legs showing (mixed-up twins).
- There is an obviously abnormal presentation (such as a head protruding with no feet).
- The lamb is showing but the ewe isn't making progress.
- The ewe is obviously becoming weak and tired, and nothing seems to be changing.
- The ewe has been in obvious labor for a couple of hours with no sign of change.

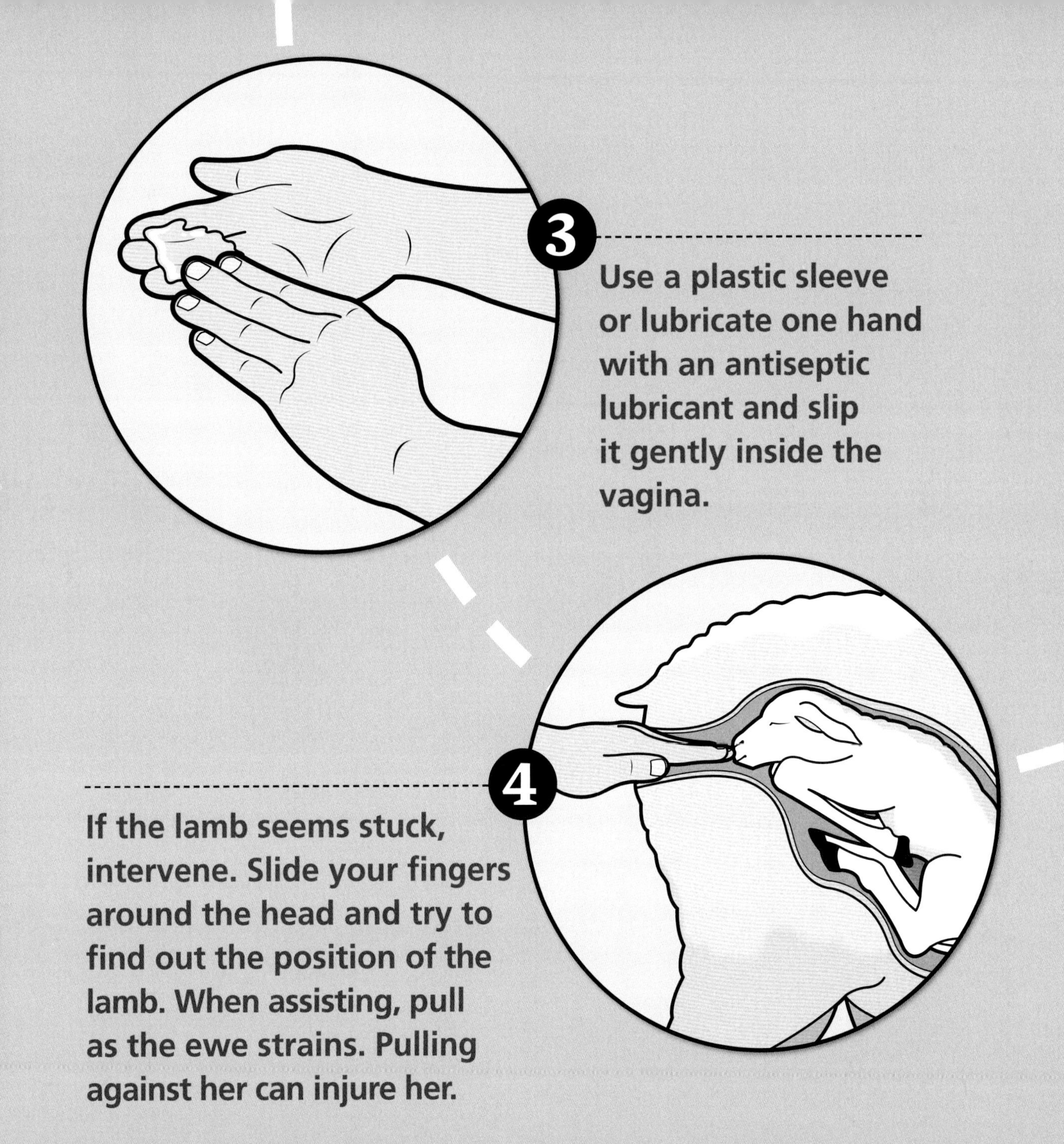

**3** Use a plastic sleeve or lubricate one hand with an antiseptic lubricant and slip it gently inside the vagina.

**4** If the lamb seems stuck, intervene. Slide your fingers around the head and try to find out the position of the lamb. When assisting, pull as the ewe strains. Pulling against her can injure her.

# ABNORMAL BIRTHS

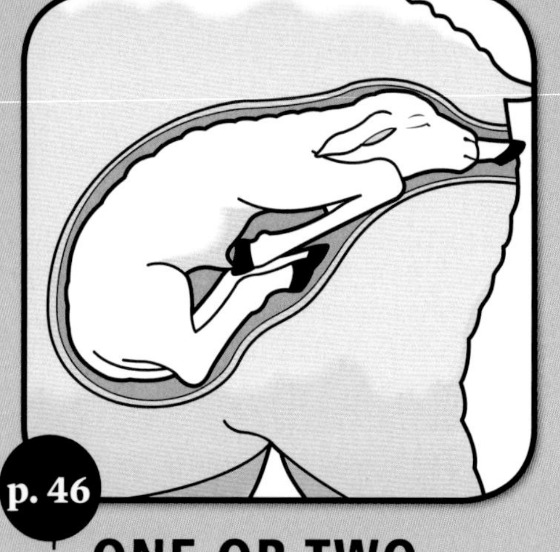

p. 46

### ONE OR TWO LEGS BACK

Either one or both of the legs are bent back. You will need to use a lamb puller.

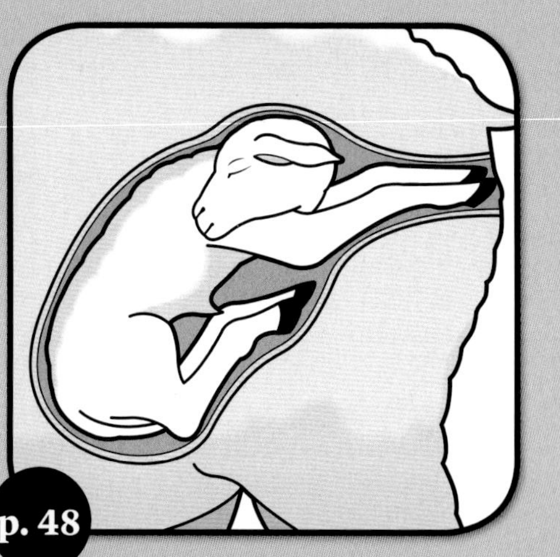

p. 48

### HEAD TURNED BACK

Both legs are presenting; the head may be turned to one side or down between the front legs. This requires emergency veterinary attention.

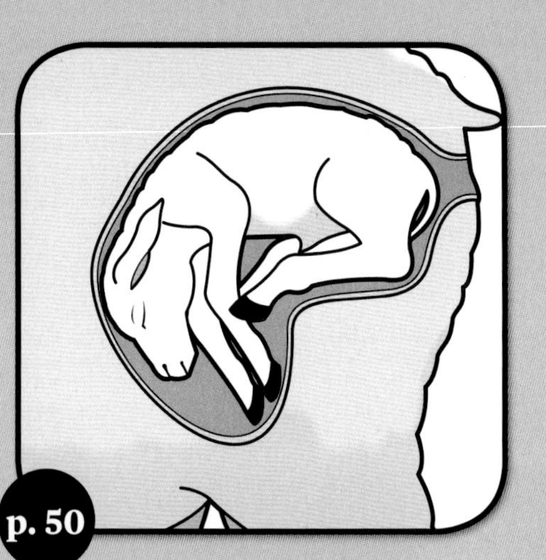

p. 50

### BREECH / HIND FEET FIRST

The lamb presents backward, its tail toward the pelvic opening and the hind legs pointed away from the opening.

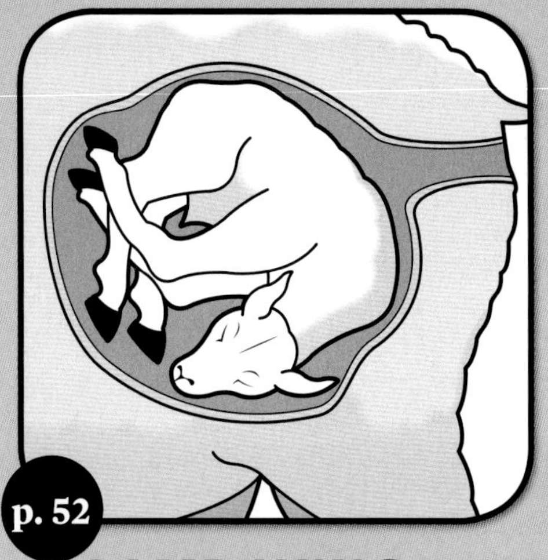

p. 52

### LAMB LYING CROSSWISE

The lamb lies across the pelvic opening and only the back can be felt.

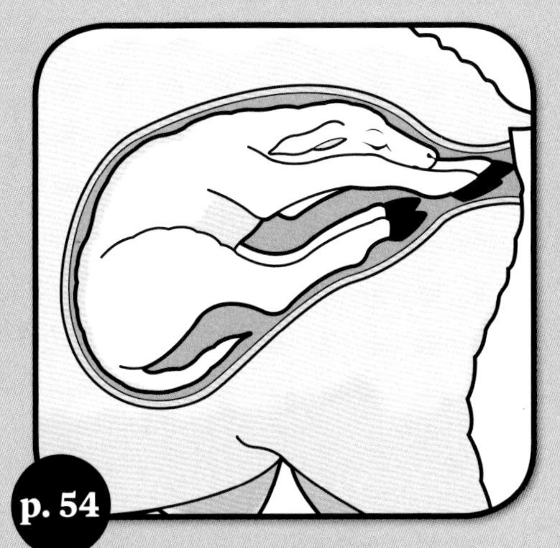

p. 54

## FOUR LEGS PRESENTING

All four legs are entering the birth canal at once.

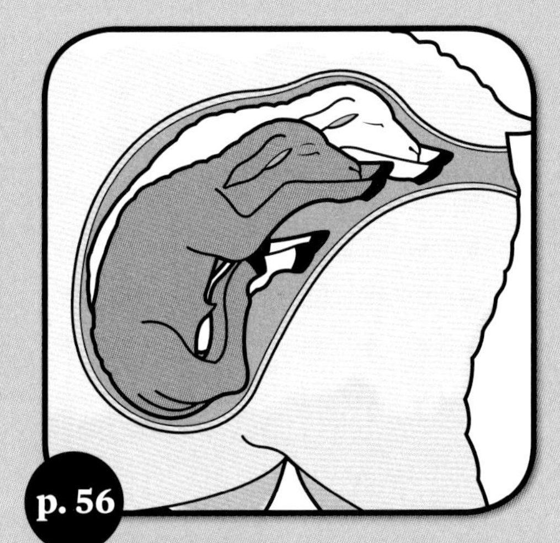

p. 56

## TWINS TOGETHER

Both lambs are presenting normally, but at the same time.

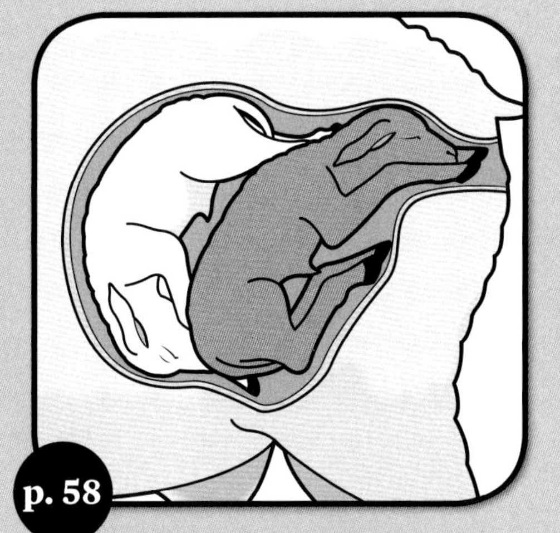

p. 58

## TWINS, ONE BACKWARD

One lamb is presenting normally and the other is coming hind feet first.

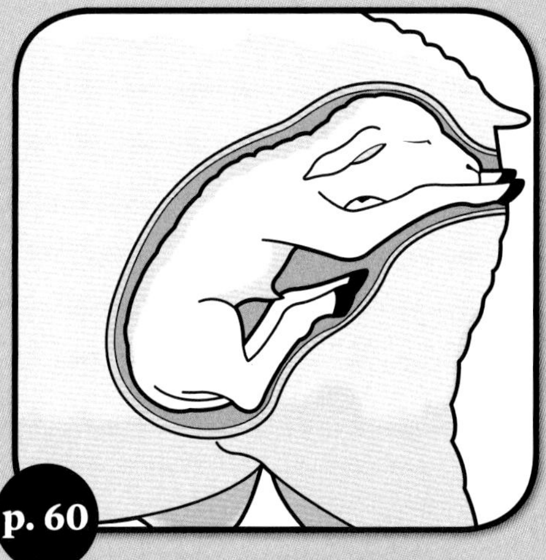

p. 60

## LARGE HEAD OR SHOULDERS

Lamb is presenting normally but its head and shoulders are too large for the pelvic opening.

# One or Two Legs Back

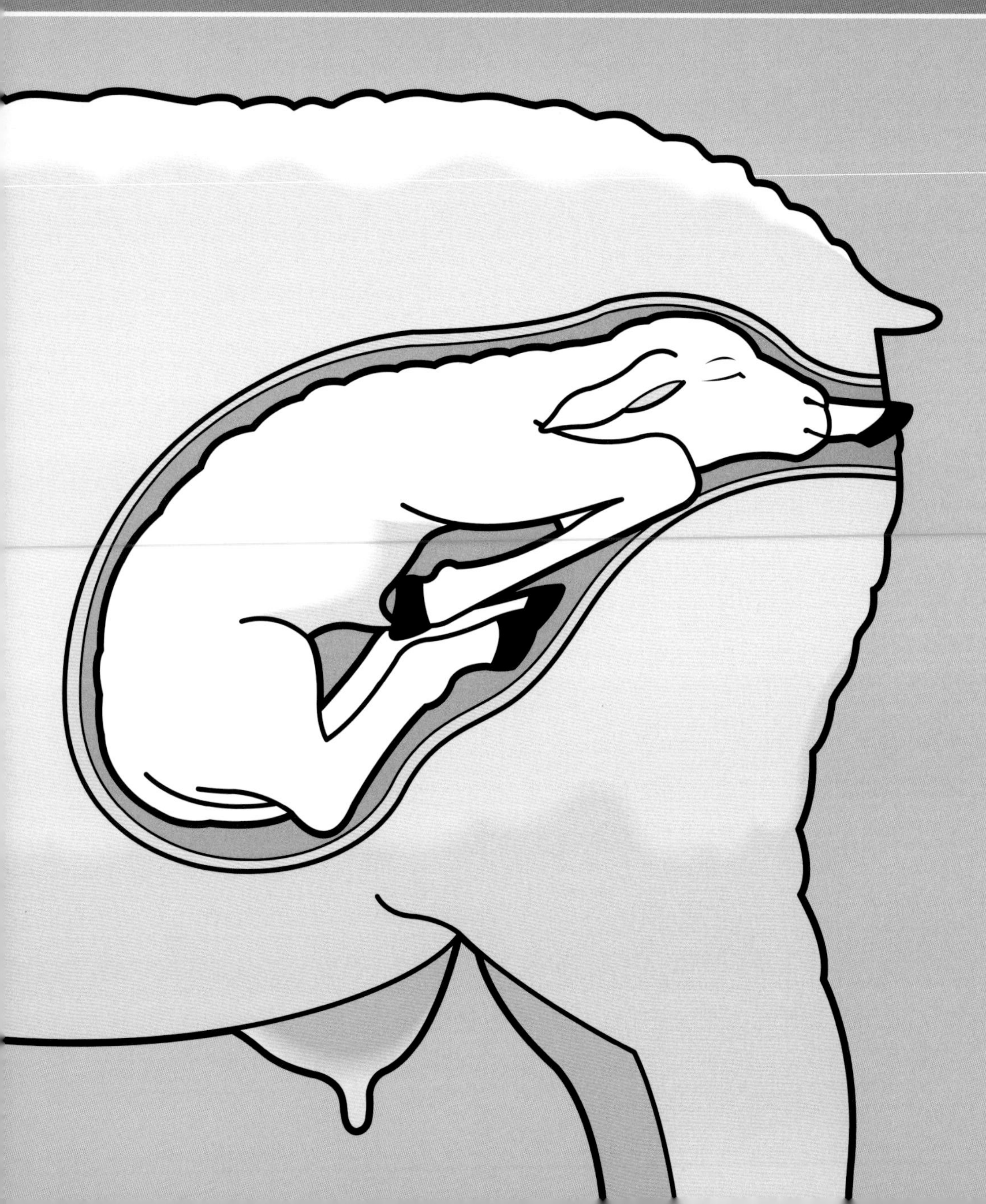

**YOU WILL NEED A LAMB PULLER** to deliver a lamb in either of these positions. The lamb puller will help you to bring the retained leg forward so that you can pull out the lamb in normal position. The cord on the head is important, for the head may drop out of the pelvic girdle, making it difficult to get it back in again.

The key to assisting is to get the lamb back in a normal position and then gently guide it out. If you apply force to an incorrectly positioned lamb, you will cause damage to the lamb and the ewe.

## LAMB PULLER

**You can use some twine or small ropes as a lamb puller. Simply make a slipknot with the twine and create a loop to place around the lamb's head and legs, as shown below. Commercial lamb pullers made from strong cable and covered in plastic tubing are also available. Pull in time with the ewe's contractions.**

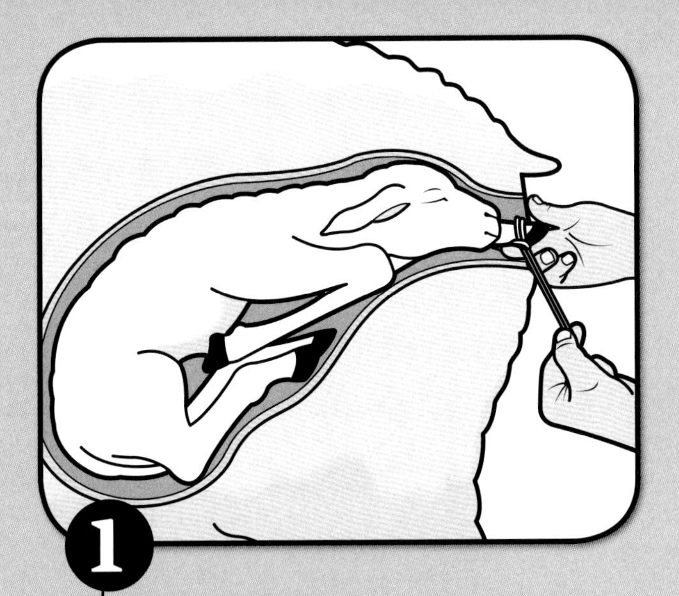

1

Lay the ewe on her side so the protruding leg is on the bottom. Slip a piece of twine around the protruding foot. If both legs are back, proceed one leg at a time.

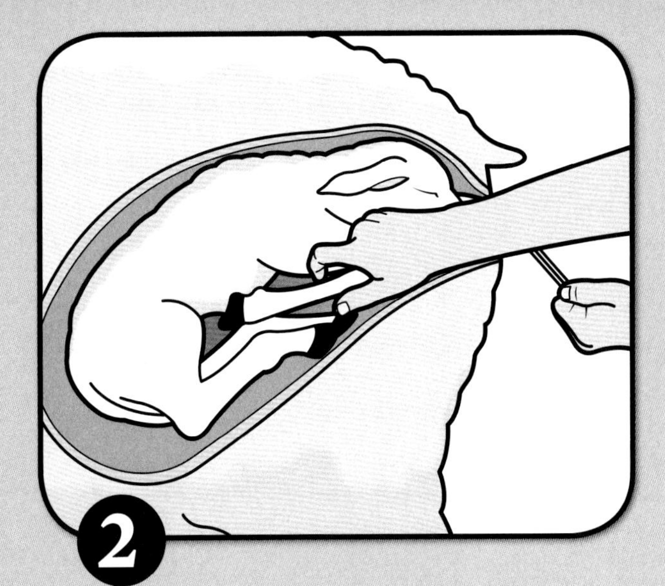

2

Gently push back the lamb as you reach into the ewe to locate the bent foreleg.

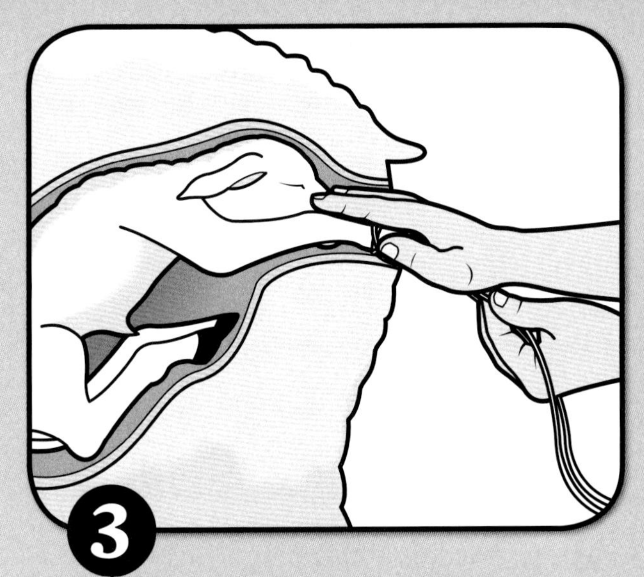

3

Draw the bent leg forward to align with the other leg.

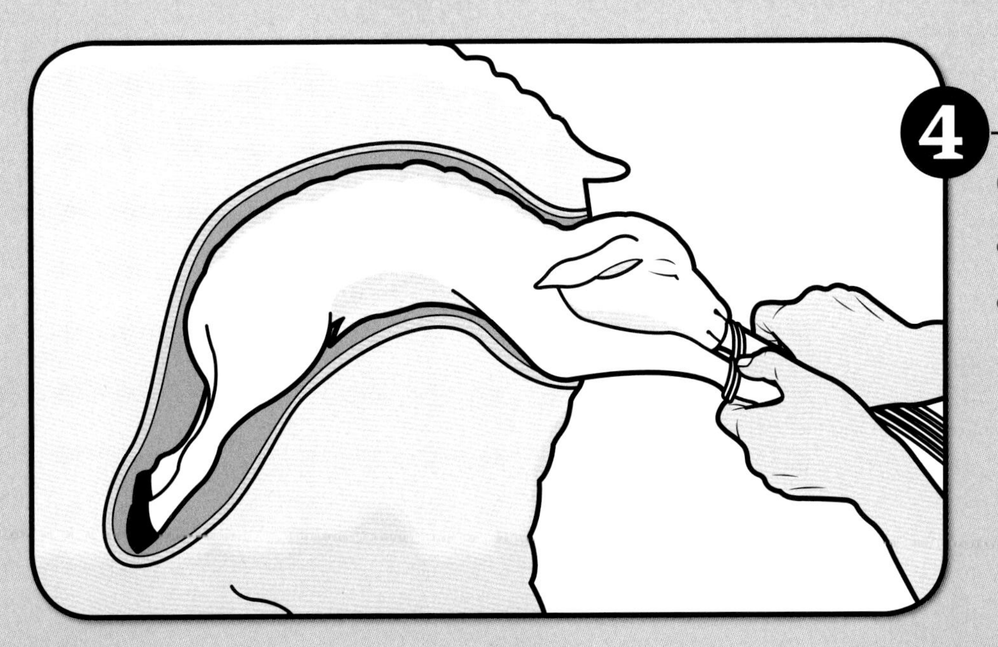

4

Grab each foot and pull out and down.

# Head Turned Back

**IN THIS PRESENTATION,** the head may be turned back to one side along the lamb's body or down between its front legs. This is one of the most difficult abnormal presentations to deal with and often requires a cesarean section, or surgical removal, to get the lamb out. Call the veterinarian when you first detect this presentation, so that he or she will be ready to assist in case you are unable to pull out the lamb using the method described below.

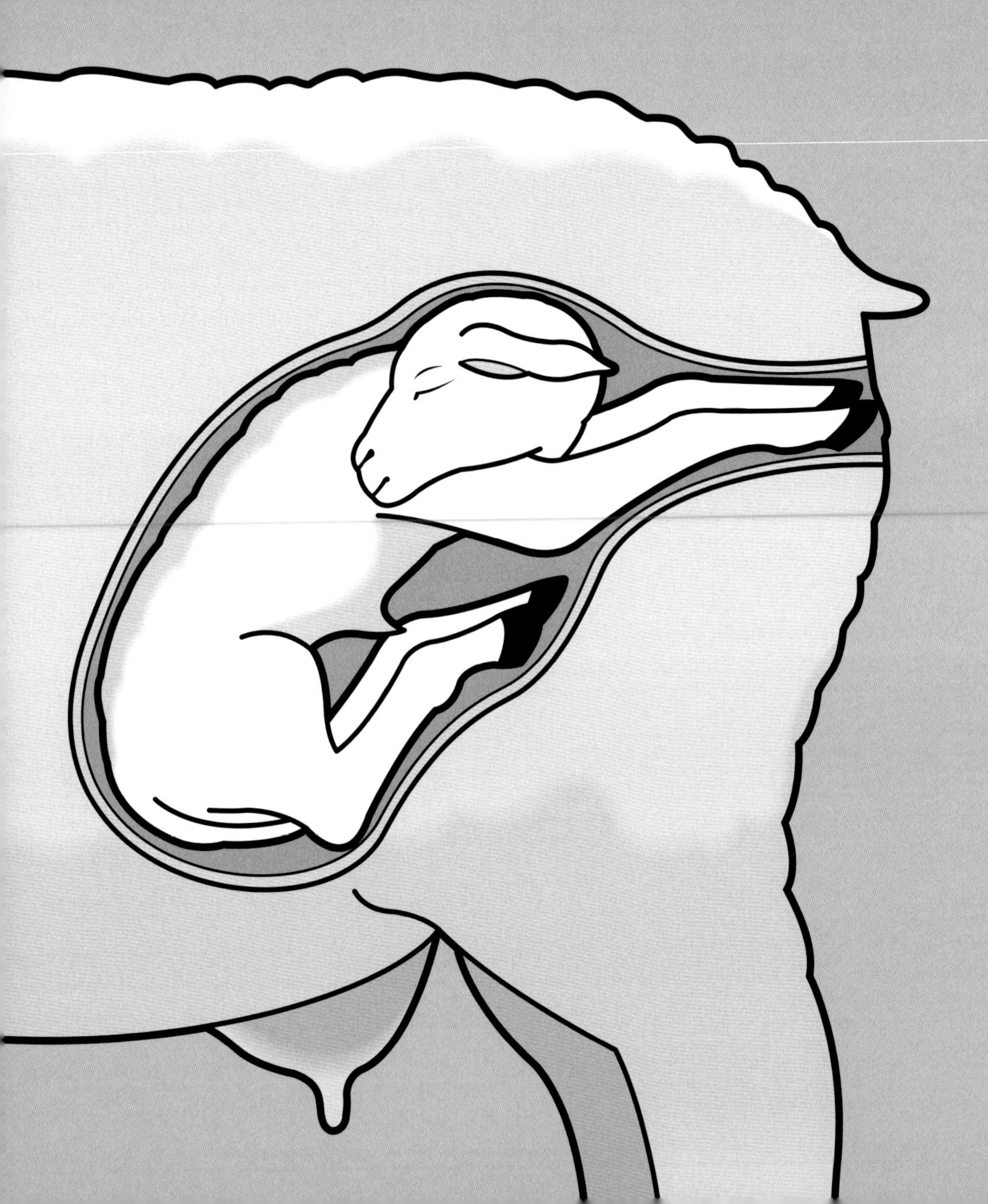

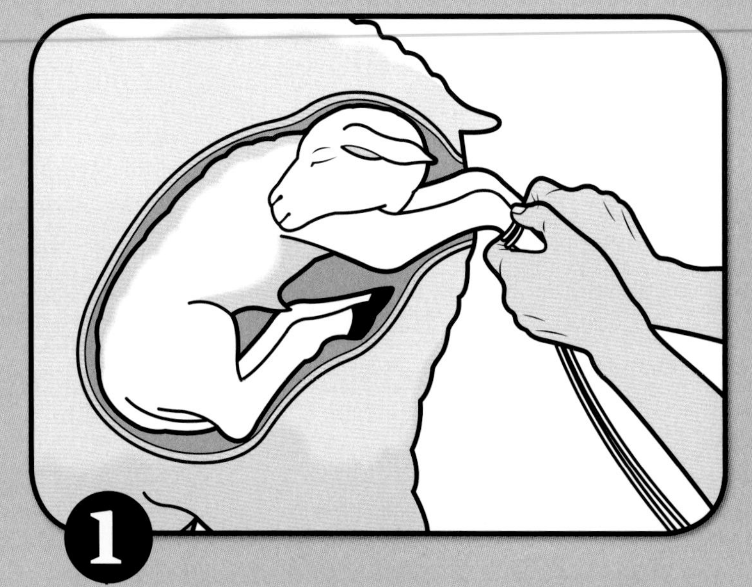

**1** If the front legs are showing, slip a noose of heavy twine over each front leg.

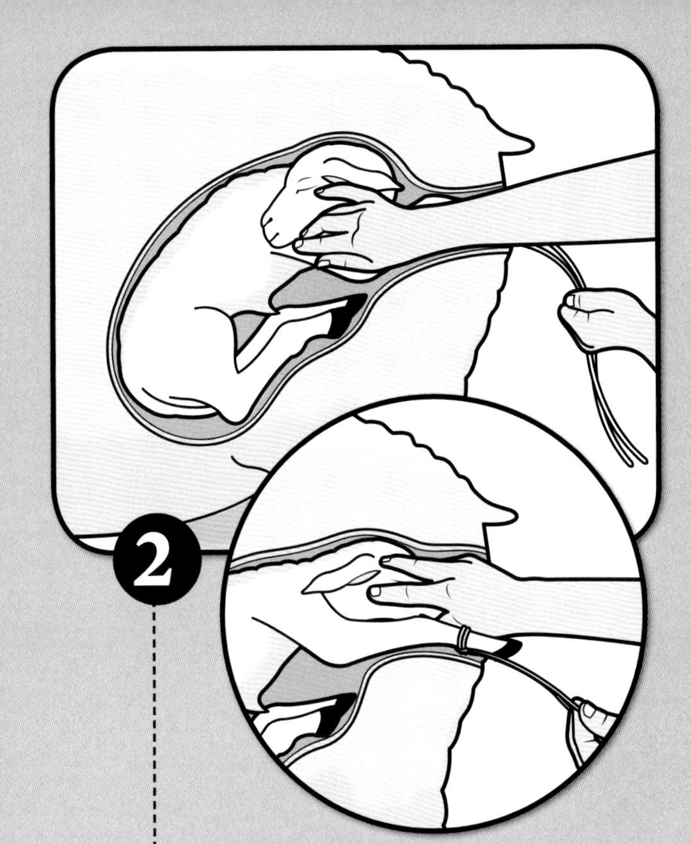

**2** Insert a lubricated hand and push the lamb back. Bring the head forward into its normal position.

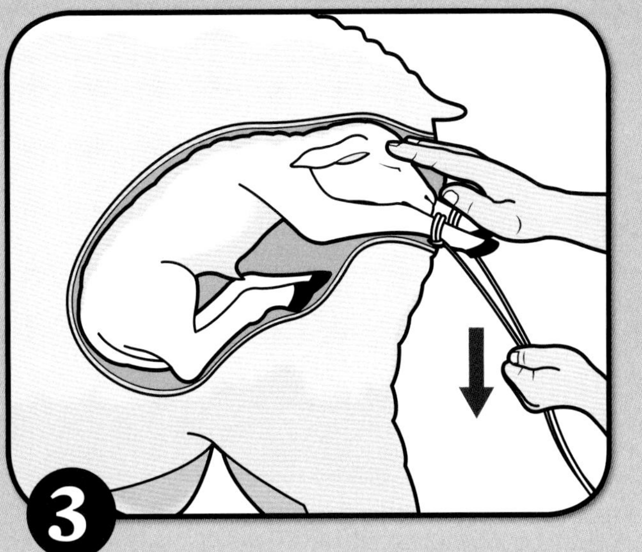

**3** While pulling the legs forcefully downward, guide the head through the pelvic opening. The head and feet should emerge at the same time.

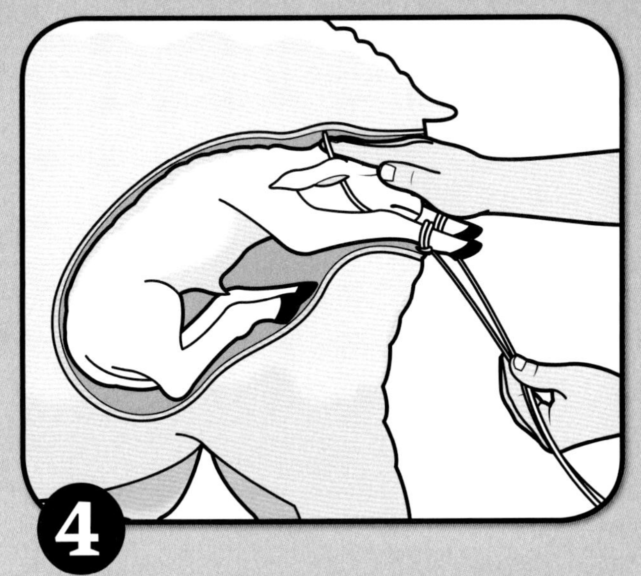

**4** If you have a hard time gripping the slippery head, attach a rope around the lamb's head, behind the ears. Pull on the rope to guide the head into position.

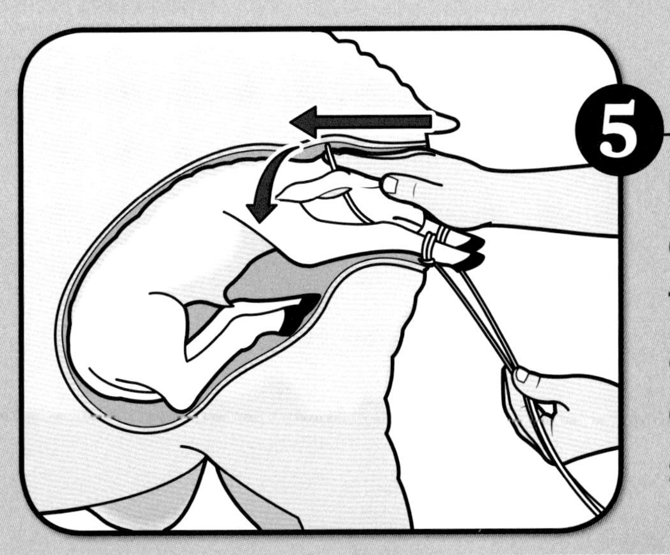

**5** If the head does not come out easily, with the cords still attached to the legs, push the lamb back again and pull one leg in front of the other to gently rotate the shoulder a half turn.

# BREECH/HIND FEET FIRST

**USE THESE STEPS FOR A BREECH POSITION.** If hind feet are already in the birth canal, follow steps 3 and 4. If after an hour or two the ewe is too exhausted to continue, determine if there is a second lamb. If not, give her a penicillin shot or insert a bolus to prevent infection, and call the veterinarian.

**TIP**

Slightly elevate the ewe's hind end (use a hay bale or metal garbage can).

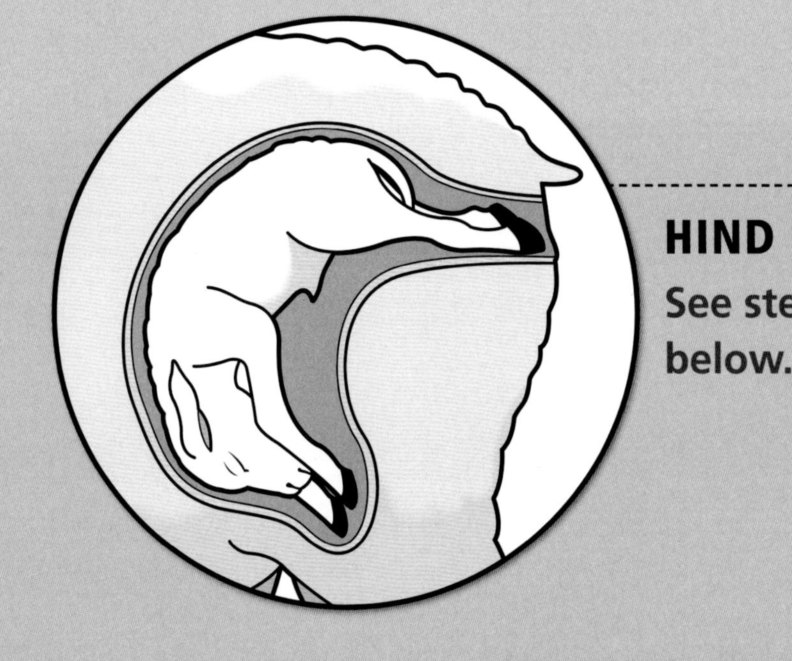

**HIND FEET FIRST**

**See steps 3 and 4 below.**

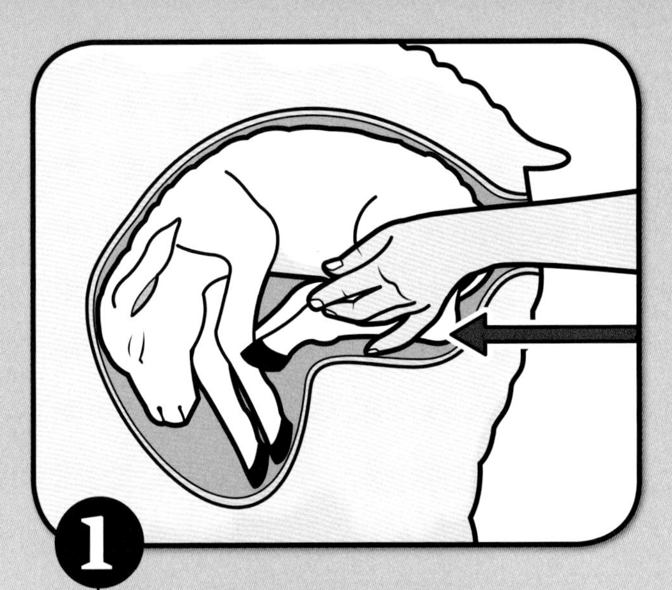

1

Push the lamb forward in the womb. Slip your hand under the lamb's rear.

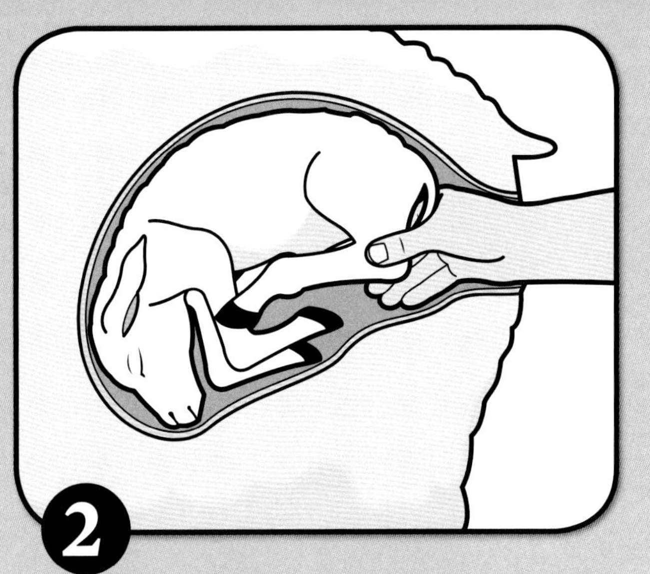

2

Take the hind legs, one at a time, flex them, and bring each foot around into the birth canal.

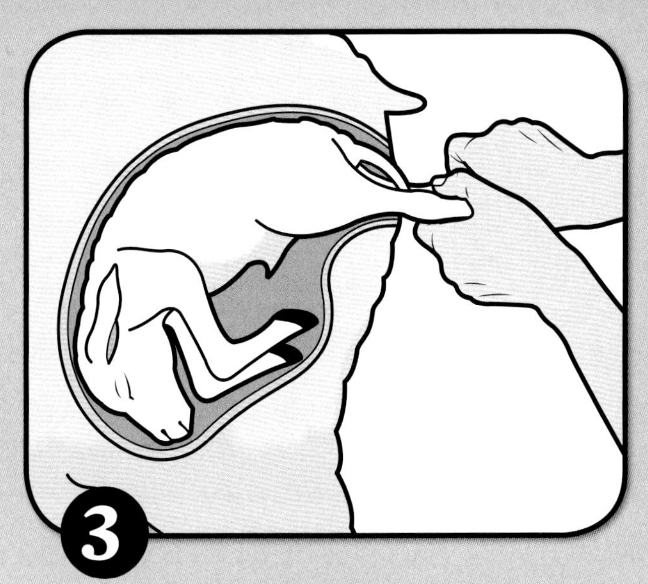

3

Turn the lamb a quarter turn and pull both legs gently until the ribs are out. Swing the lamb from side to side if it gets stuck.

4

Past the ribs, pull it downward and out quickly (it is extremely important to pull quickly!).

# LAMB LYING CROSSWISE

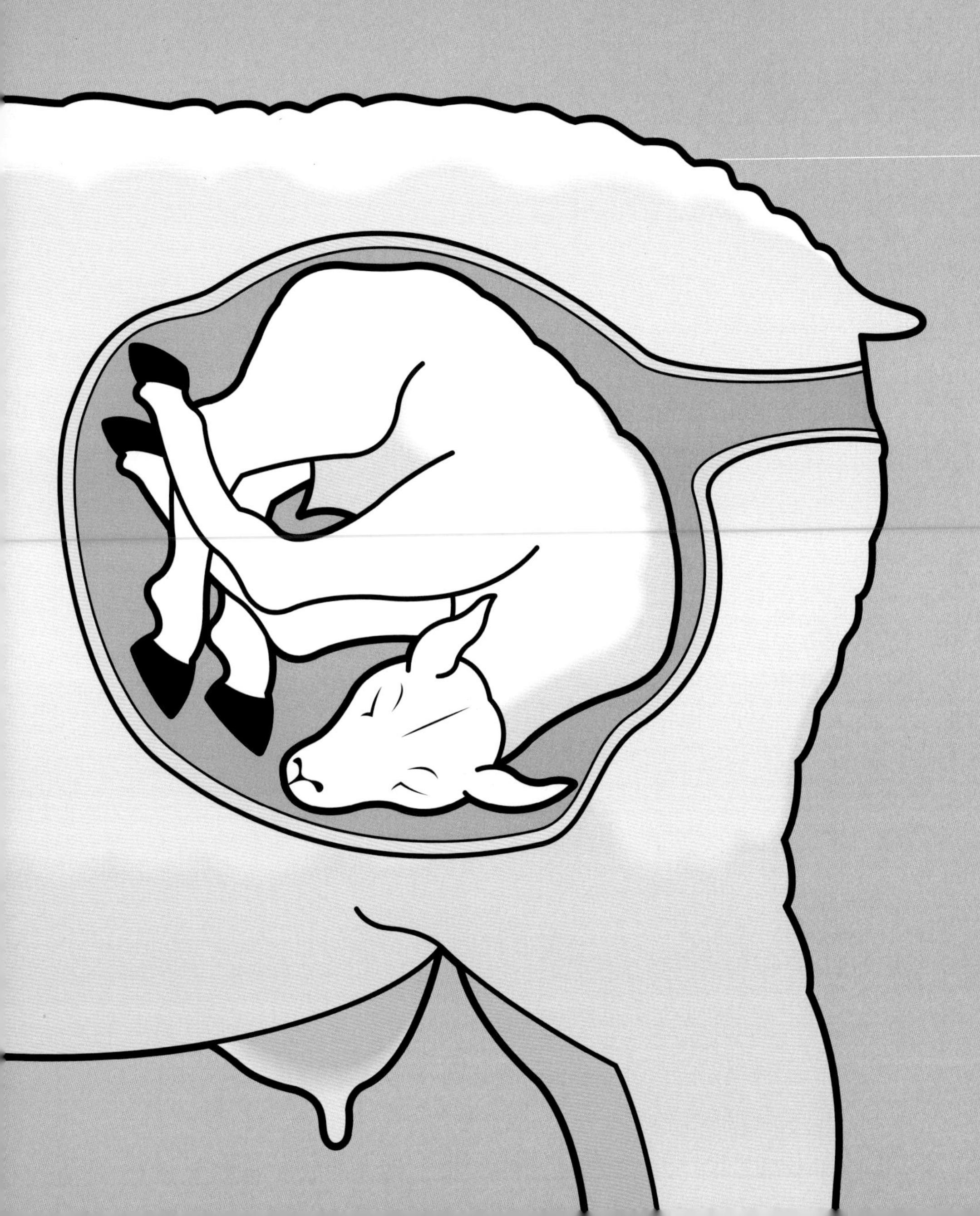

**SOMETIMES A LAMB LIES ACROSS** the pelvic opening and only the back can be felt. In some cases, you may be able to turn the lamb around to a normal position and deliver it. Otherwise, you may need to deliver it by the hind legs. In either case, follow the steps below to deliver the lamb.

## ALWAYS CHECK EYES

**Check the eyes of all newborn lambs at birth. Sometimes the lower eyelid, or both, is rolled inward; this hereditary disorder is called entropion. When it happens, the eyelashes chafe the eyeball, causing the eye to water constantly, inviting infection and even blindness. The easiest method of correcting entropion is to inject 1 mL of penicillin just under the skin beneath the lower eyelid, which forces it down into its correct position.**

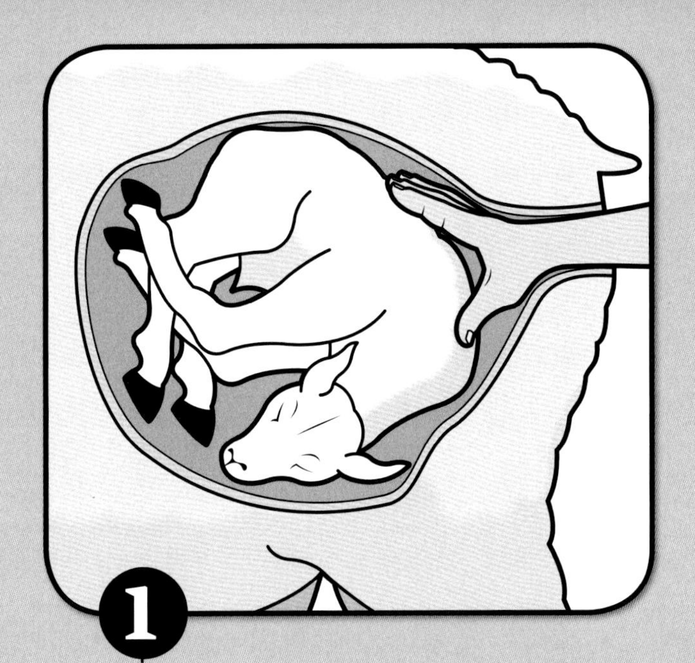

**1**

Push the lamb back a little and feel in which direction it is lying.

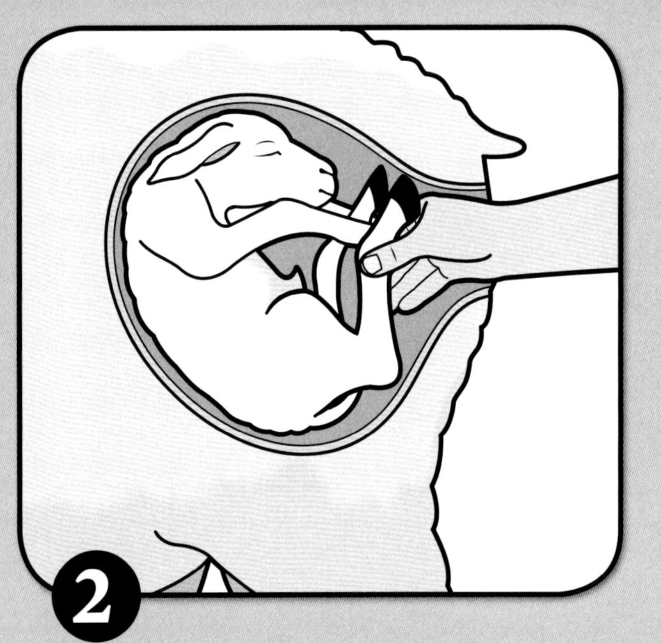

**2**

If the hind feet are close to the opening, pull the lamb out hind feet first. You will need to turn the lamb over so that its spine is in alignment with the ewe's spine.

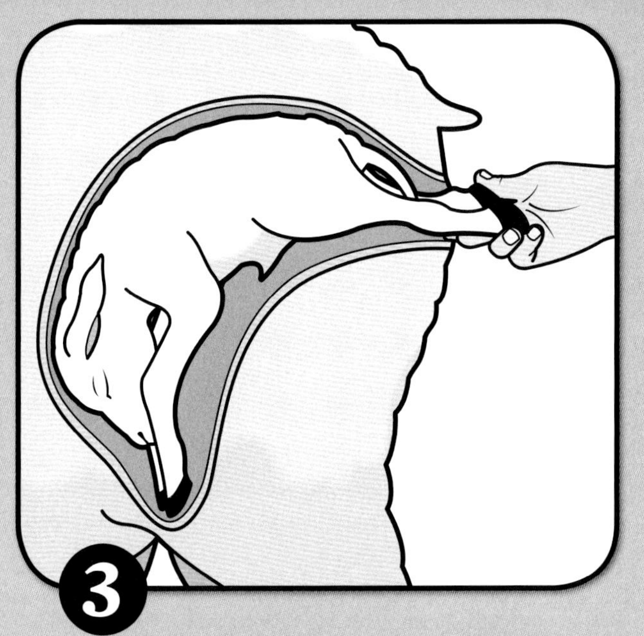

**3**

If the lamb is upside down, turn it a half turn and deliver.

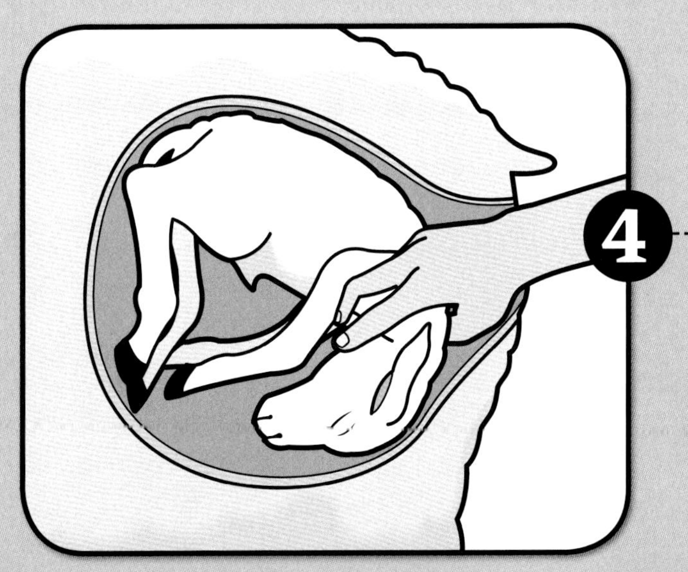

**4**

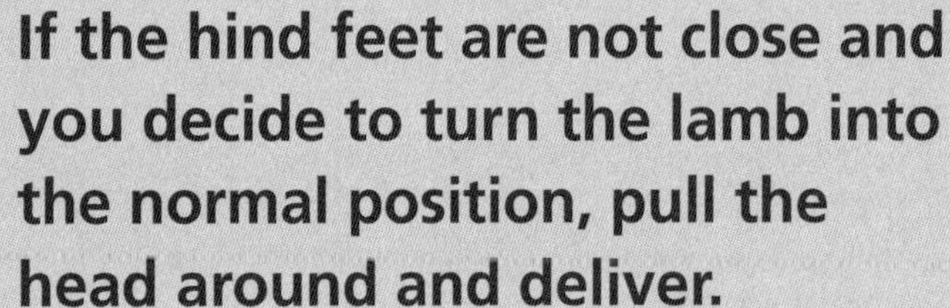

If the hind feet are not close and you decide to turn the lamb into the normal position, pull the head around and deliver.

# Four Legs Presenting

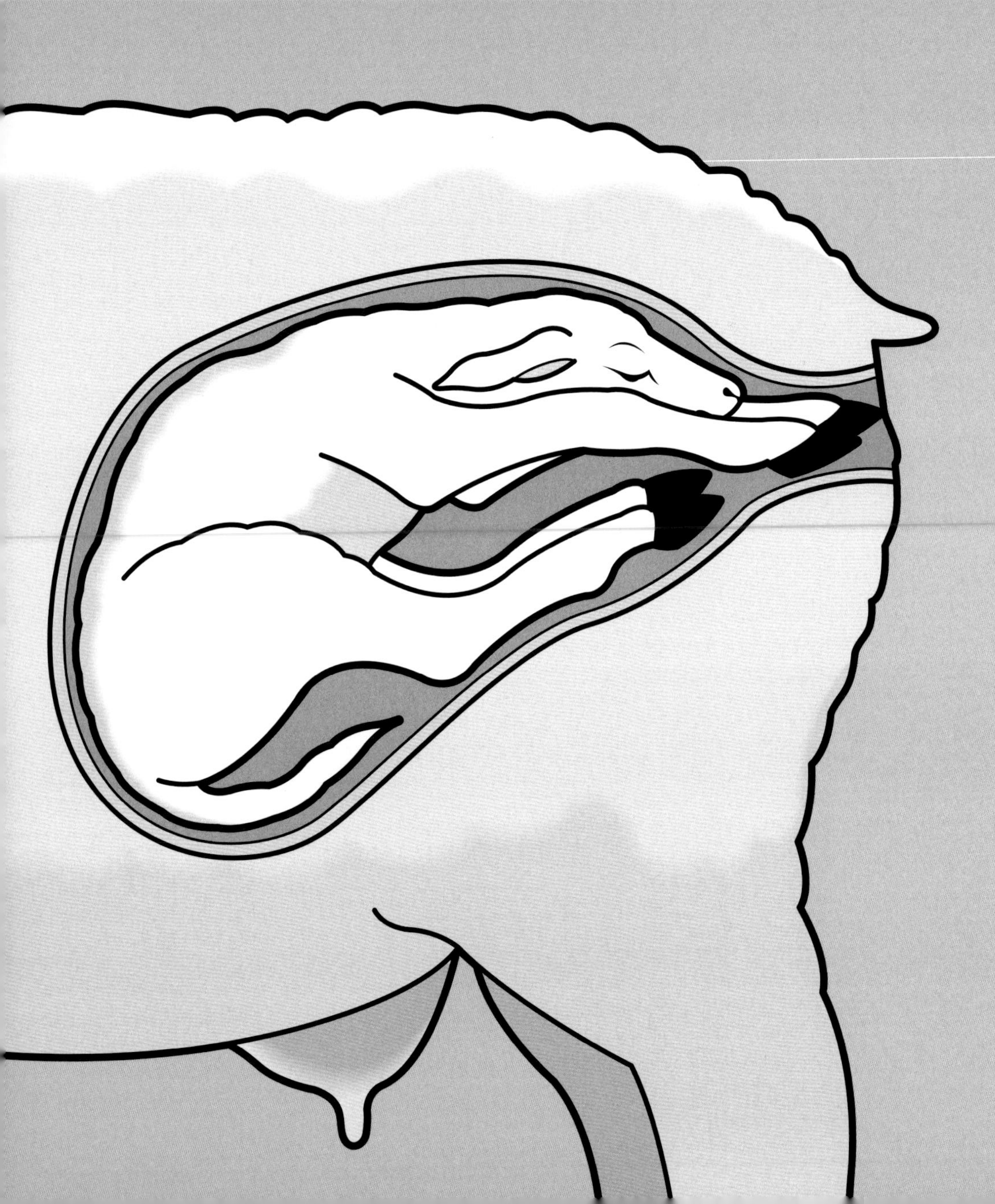

**IN THIS CASE,** you will be able to feel four legs close to the birth canal. If the head is out of position, deliver by the hind legs so you don't have to reposition the head. If you choose the front legs, the head and legs must be maneuvered into the correct birthing position. Be sure to attach the cords to the legs before pushing the lamb back to position the head, so you don't "lose" them.

## LAMBING CHECK

**Do the following as lambing time approaches.**

- **Stop, look, and listen for lambing activity.**
- **Observe quietly from a distance; look for ewes isolating themselves.**
- **Check the remote corners of the field.**
- **Keep dogs and visitors away.**
- **When checking at night, minimize light by using a flashlight.**
- **Also check ewes that have lambed. Are any lambs lost? Are they nursing?**
- **Follow the same routine each time so that ewes get used to it and will not be stressed.**

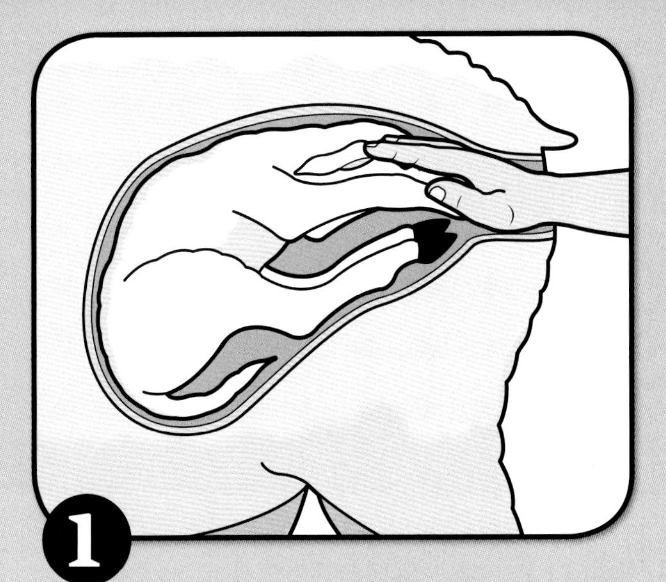


If the front legs are closer, maneuver the head into the normal birthing position (page 40) along with the legs and deliver.

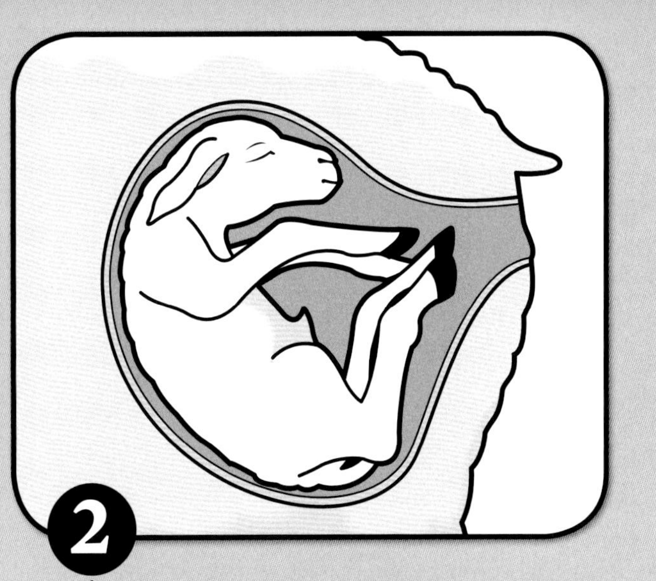


If the hind legs can be reached as easily as the front, deliver by them (shown here).

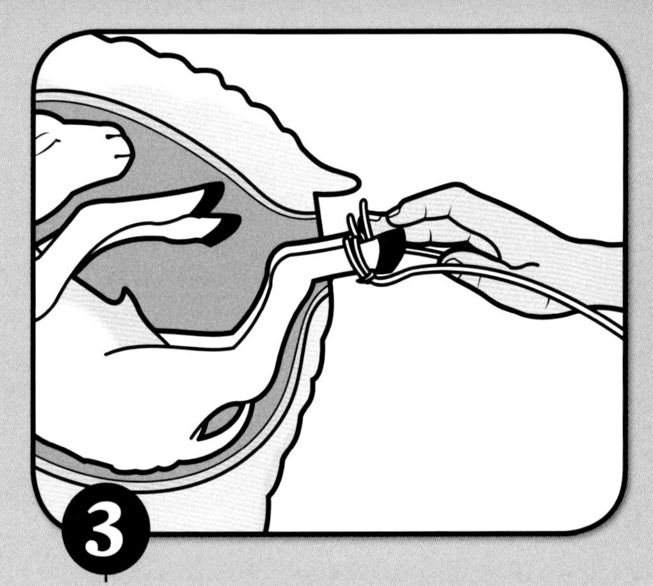


Attach cords one at a time to the legs you are working with (hind shown here) as they protrude. Push feet back in once cords are attached.

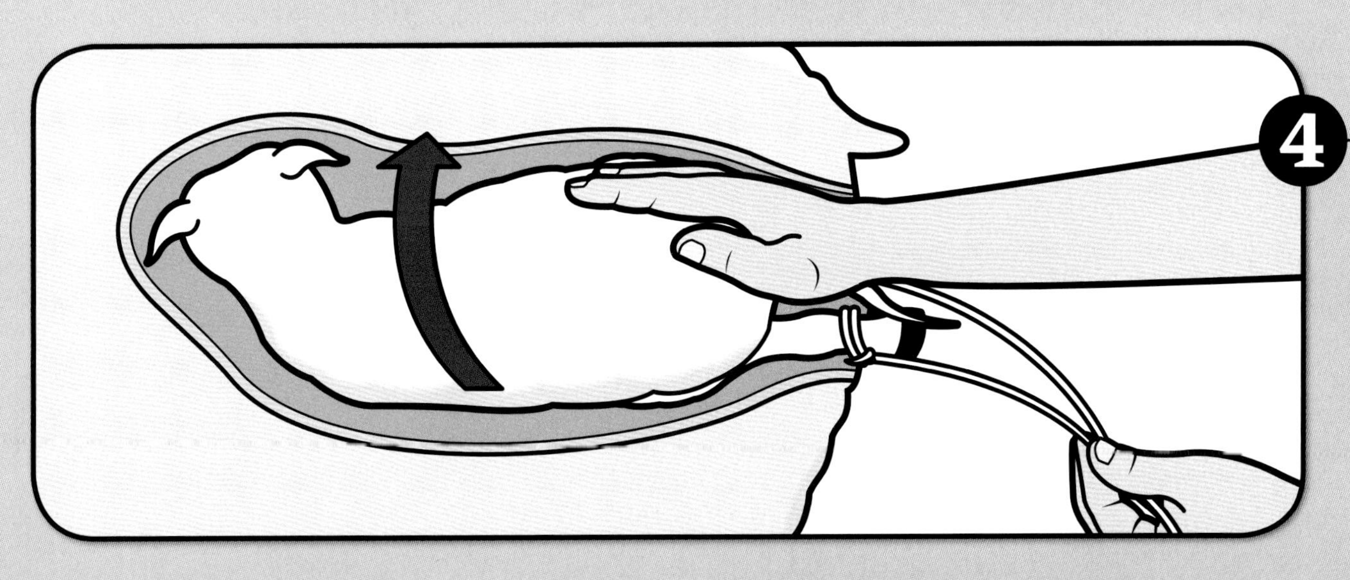

4

Give the lamb a half turn as you pull it gently downward and out.

# Twins Together

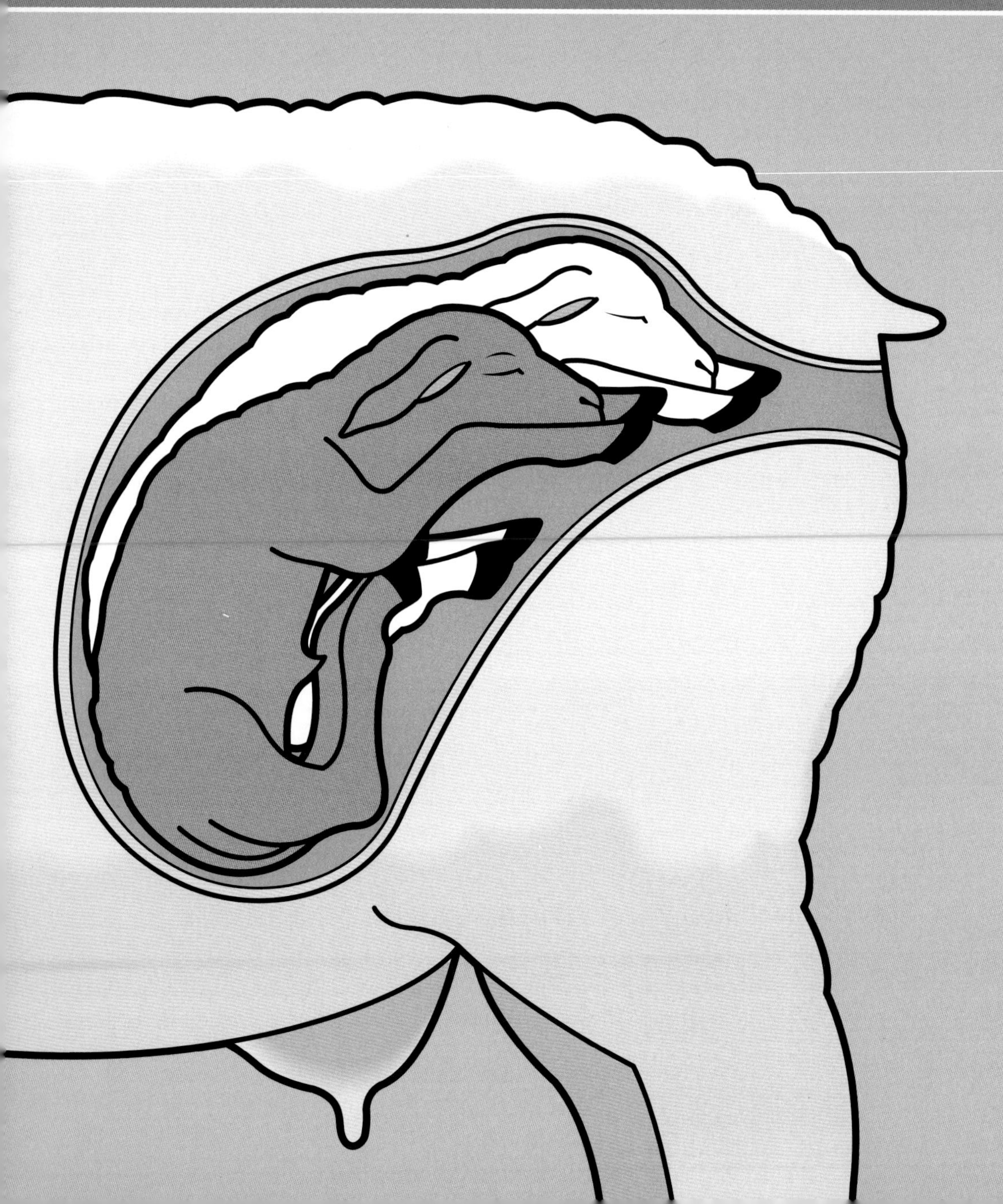

**WHEN YOU HAVE TOO MANY FEET** in the birth canal, try to sort them out. Work on delivering one lamb at a time. Trace the legs back to the body to make sure they are part of the same lamb, then position the head before pulling. You may need to push the second lamb back a little to make room for delivery of the first one.

## DEALING WITH MULTIPLE LAMBS

**With twins, triplets, and larger "litters," be sure that the ewe claims all the lambs and that they all get their share of colostrum. If the ewe does not have plenty of milk for all the lambs, start feeding her grain the day she gives birth and increase her grain consumption gradually. If you have multiple lambs that are crying a lot, they are probably not getting enough milk. You may need to give them a supplemental bottle. (See page 72 for feeding schedule.)**

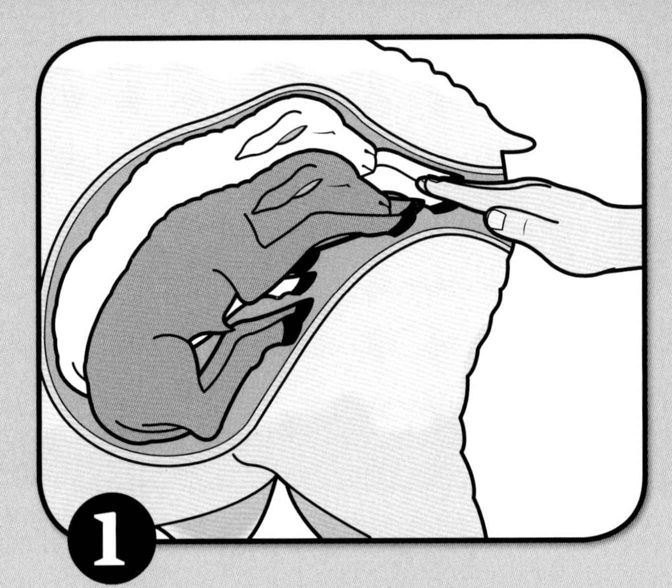

1

Try to sort out the legs in the birth canal. Work one lamb at a time.

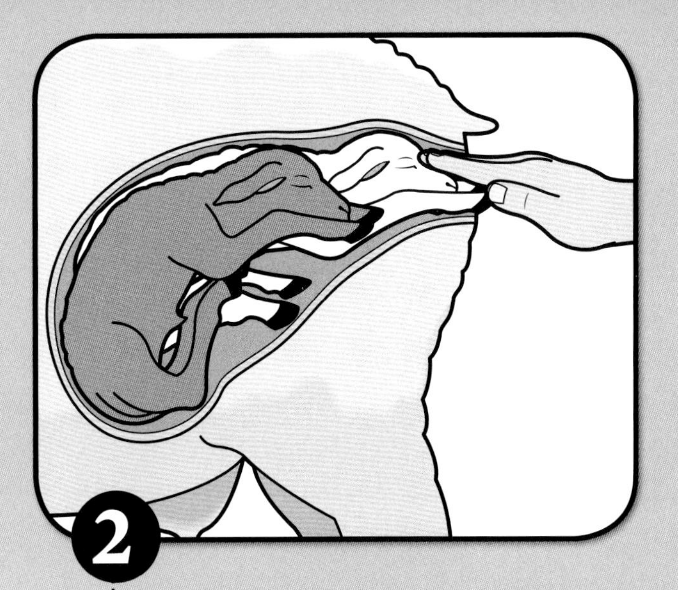

2

Feel the head to make sure it is in the right position (facing toward you). If it is not, correct the position to match the illustration above.

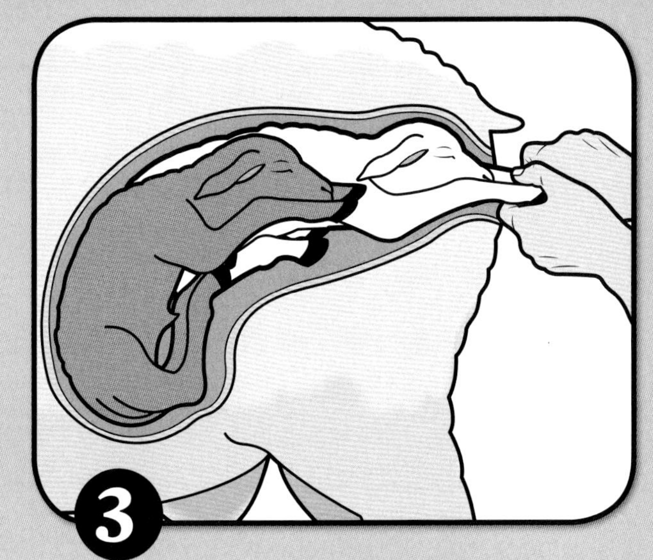

3

Pull each foot of the closer lamb toward you (trace the legs back to the body to make sure they are part of the same lamb).

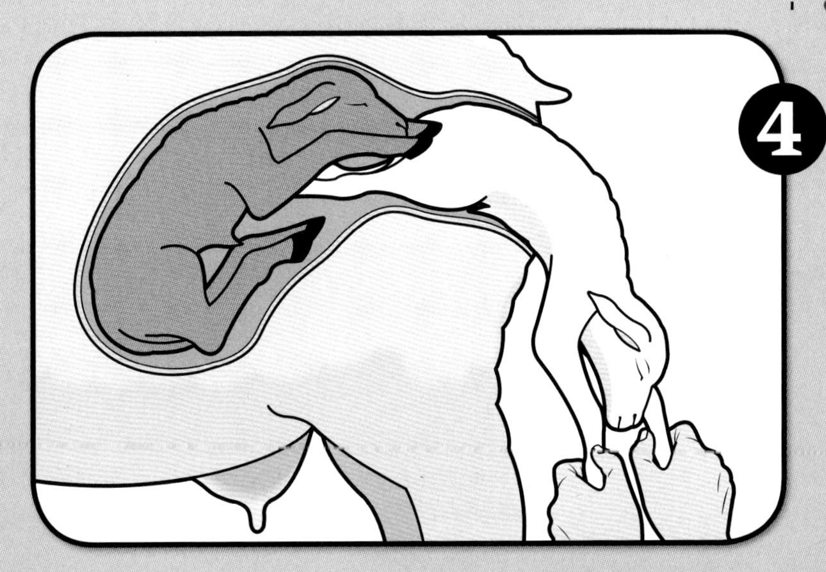

4

Deliver the first lamb. The second lamb will usually come out on its own about 5 minutes later. If the ewe labored hard with the first, however, you may need to assist.

# Twins, One Backward

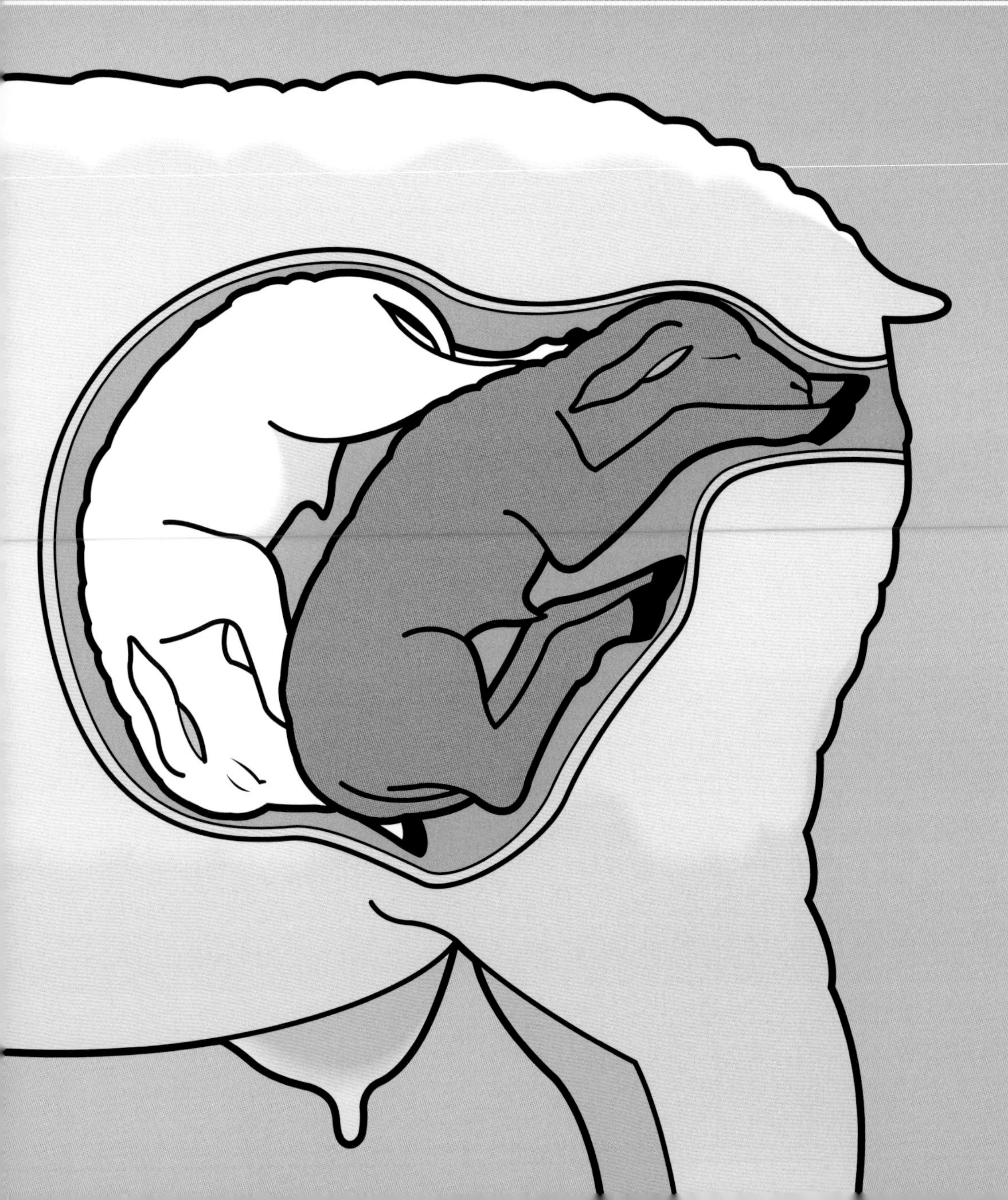

**WHEN TWINS ARE COMING OUT TOGETHER,** one and sometimes both may be reversed. It is often easier to first pull out the one that is reversed. If both are reversed, pull the lamb that is closer to the opening. Very rarely, the head of one twin is presented between the forelegs of the other twin, a confusing situation that can be sorted out by tracing the entire body of one lamb.

## CHANCES OF TWINNING

**A ewe is more likely to have twins if she and the ram she has mated with share a genetic inclination toward twinning, and if she has gone through flushing. Flushing is the practice of placing the ewe on an increasing plane of nutrition to prepare for breeding. With better nourishment, the ewe is more likely to drop two eggs.**

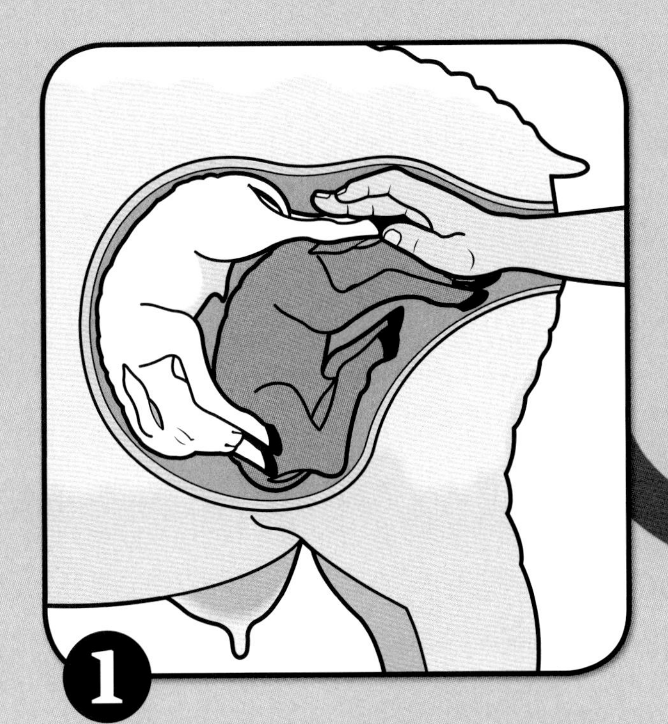

1

Determine if either twin is reversed. Follow the hind legs up to the hock and tail and follow the front legs up to the head.

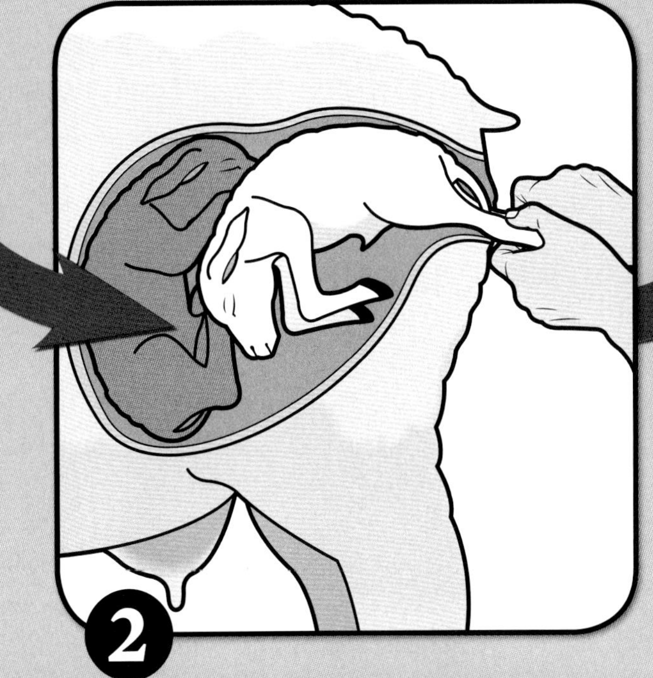

2

First pull out the one that is reversed (see page 50 for breech and hind feet first).

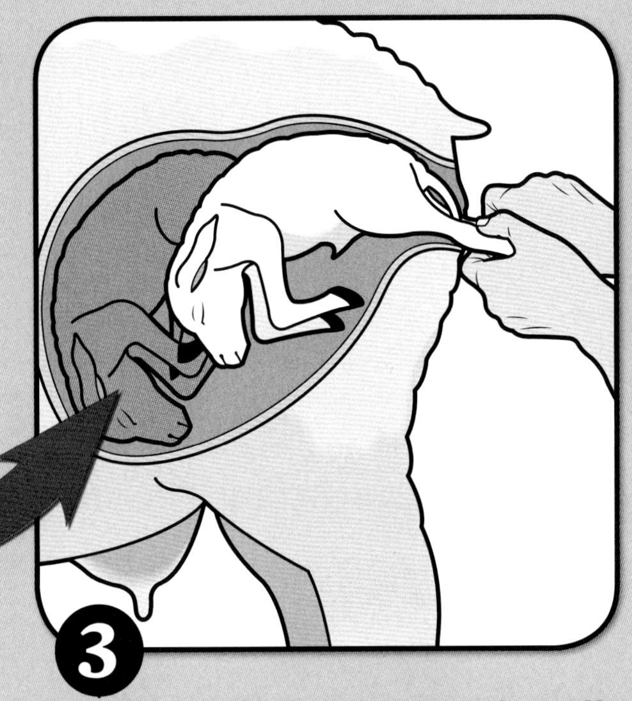

3

If both are reversed, pull out the lamb that is closer to the opening.

# Large Head or Shoulders

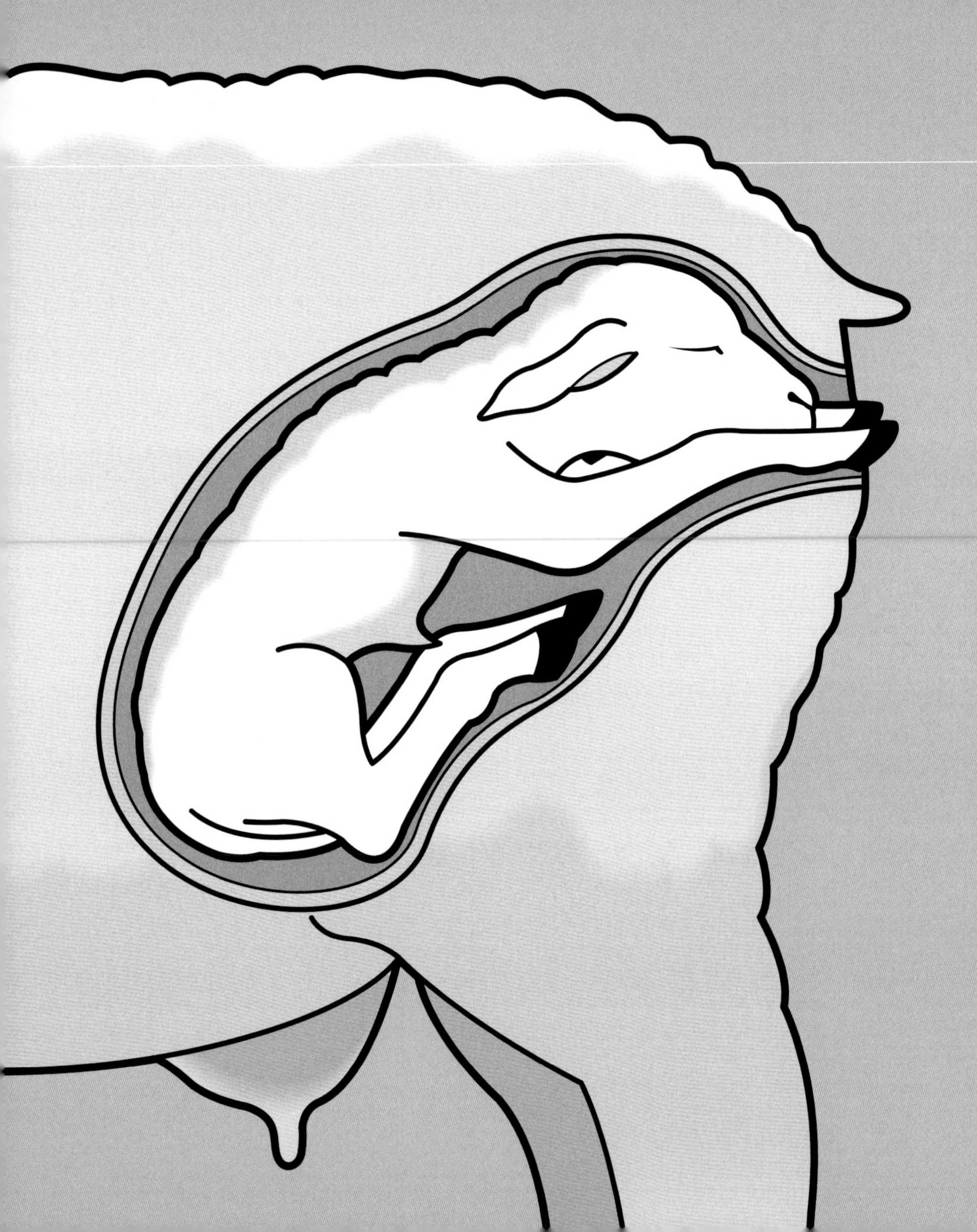

**EVEN IF THE LAMB IS IN THE NORMAL POSITION,** it may need some help if it is extra large or the ewe has a small pelvic opening. Sometimes the shoulders are large and are stopped by the pelvic opening. Occasionally, the head is large or may be swollen if the ewe has been in labor for quite a while. You should have a lamb puller and antiseptic lubricant on hand for this procedure.

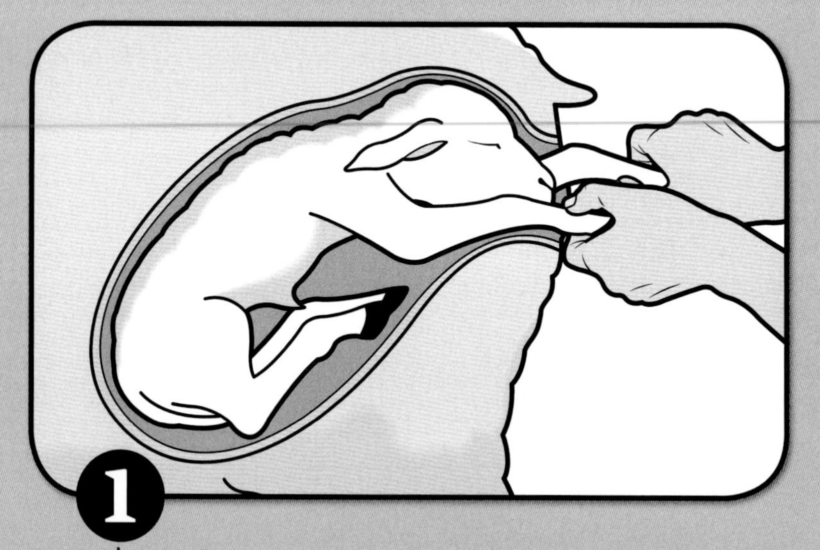

**1** **Using a gentle outward and downward pulling motion, pull to the left or right, so the shoulders go through at more of an angle and thus more easily.**

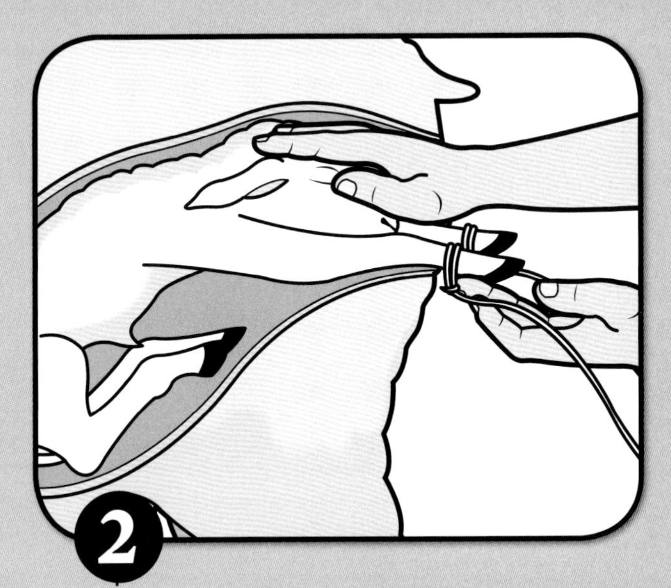


Attach a rope or snare to each leg to help you pull the lamb. Gently push back the lamb.

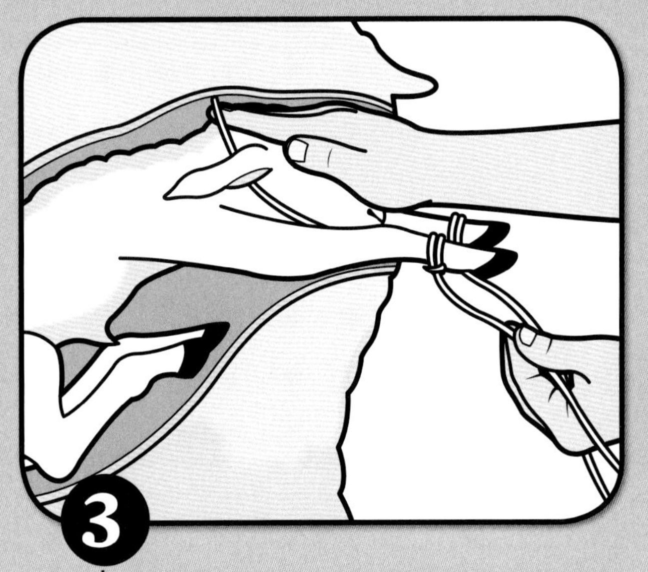


Attach a rope or snare around the lamb's head, behind the ears.

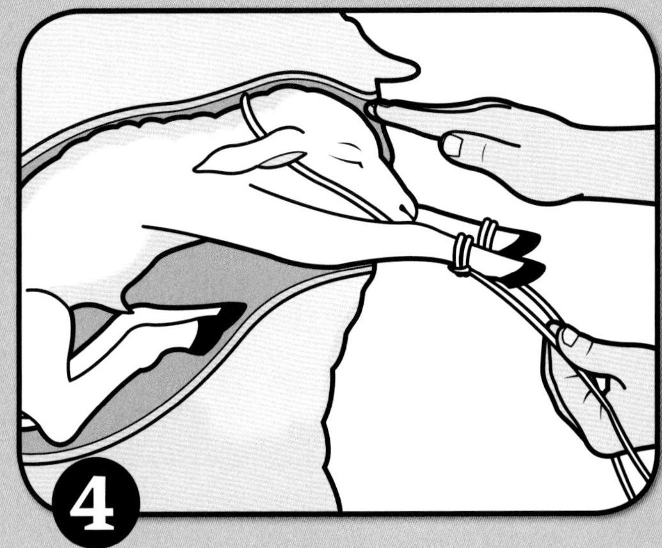


Occasionally, the head is swollen if the ewe has been in labor for quite a while. Assist by gently pushing and stretching the skin of the vulva back over the lamb's head.

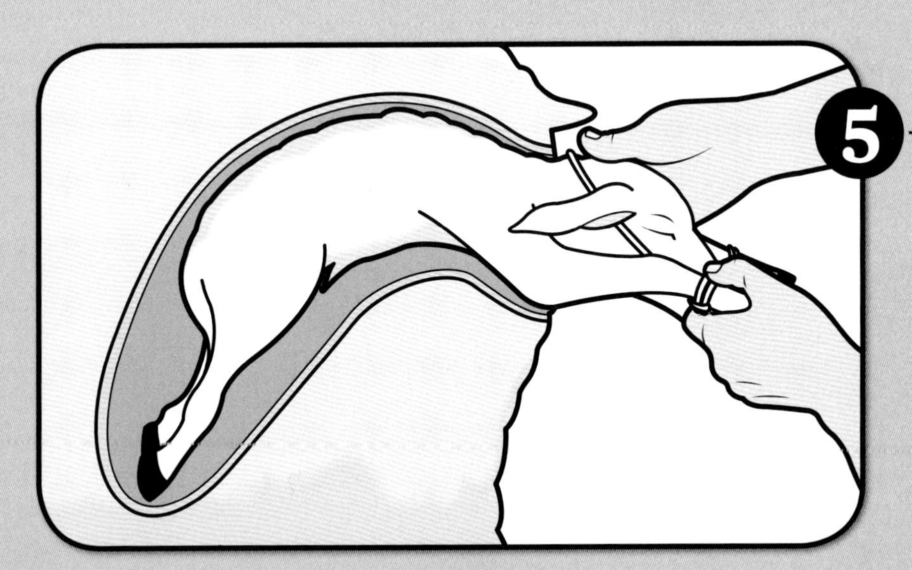

5

Once the head is through, extend the other leg completely and pull out the lamb by its legs and neck. Use mineral oil or an antiseptic lubricant with a difficult large lamb.

# AFTER LAMBING

## IMMEDIATE CARE

**1** Wipe the mucus off the lamb's nose if the ewe isn't doing it.

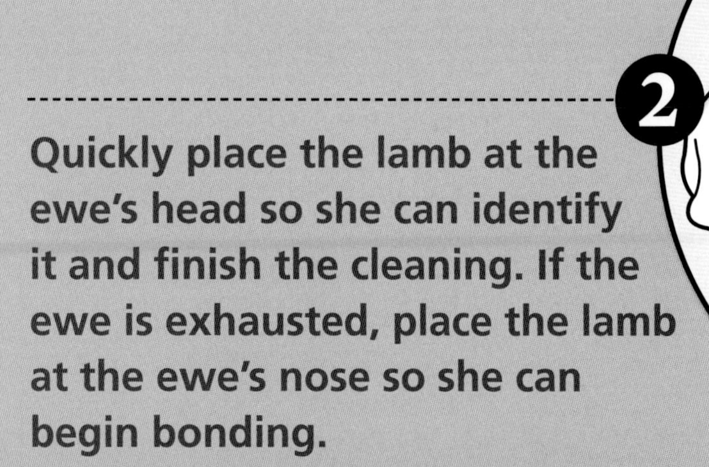

**2** Quickly place the lamb at the ewe's head so she can identify it and finish the cleaning. If the ewe is exhausted, place the lamb at the ewe's nose so she can begin bonding.

### WARMING UP A FROZEN LAMB

The warm-water method is probably the best technique for warming up a very cold lamb. Submerge it up to its neck in water that is quite warm to the touch. It's best to put a plastic bag around it so that it doesn't get wet. The lamb will begin to struggle, but keep it immersed for several minutes. Dry it well with a blow-dryer if you did not use a bag, and place the lamb in a warm environment until it has recovered fully. Feed the lamb 1 to 2 ounces (30–60 mL) of warm colostrum or milk replacer as soon as it can take it. If you are experienced, force-feeding with a stomach tube after removal from the water and drying speeds up recovery.

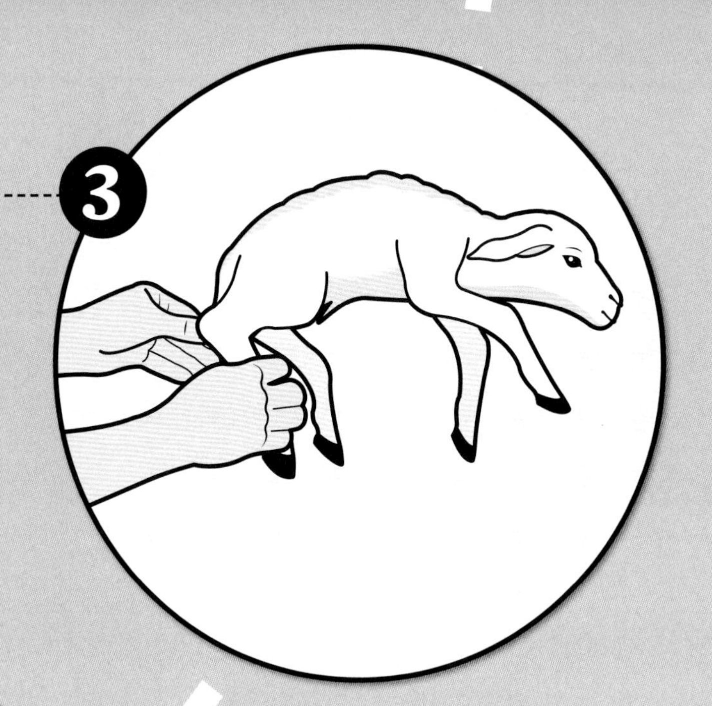

**Help dry off the lamb with old towels or old flannel sheets and be sure not to remove the lamb from its mother's sight.**

### "UNPLUG" THE EWE

If the lamb doesn't seem to be getting milk, "unplug" the ewe by breaking the waxy plug on the end of the teat and stripping it of several squirts of colostrum.If the ewe will not be able to feed the lamb, now is the time to graft it (page 70).

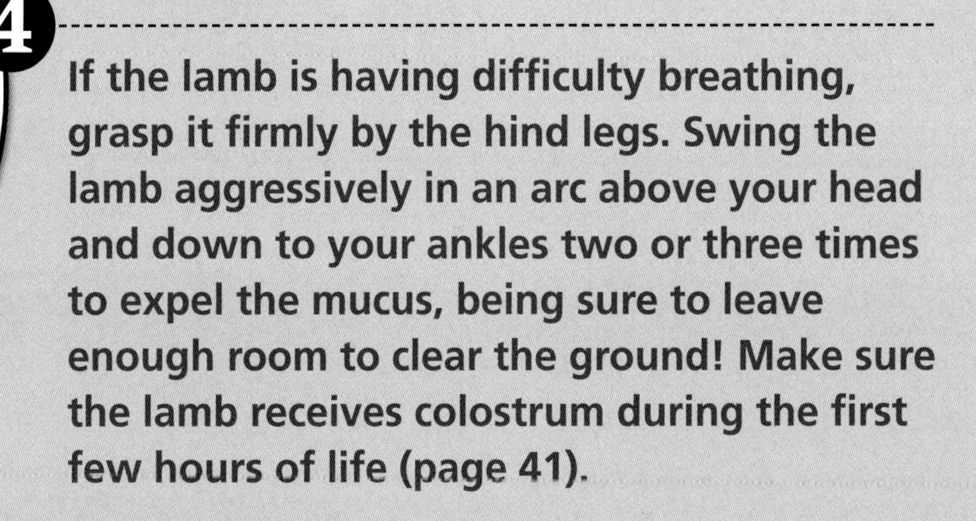

**If the lamb is having difficulty breathing, grasp it firmly by the hind legs. Swing the lamb aggressively in an arc above your head and down to your ankles two or three times to expel the mucus, being sure to leave enough room to clear the ground! Make sure the lamb receives colostrum during the first few hours of life (page 41).**

# Treating the Navel

**IODINE SHOULD BE APPLIED AS SOON AS POSSIBLE** after birth to prevent bacteria from entering the newborn lamb's navel. Avoid spilling the iodine on the lamb or applying too much of it. As an extra precaution, you can treat the cord again 12 hours after the first application.

1. Place a 7 percent tincture of iodine solution in a small plastic jar with a wide mouth.

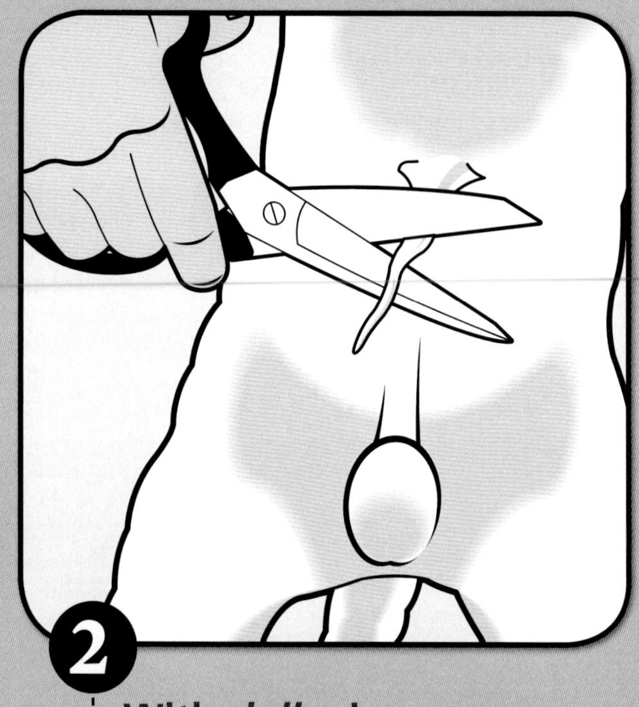

2. With *dull* scissors or a *dull* knife, snip the umbilical cord to about 2 inches (5 cm) long.

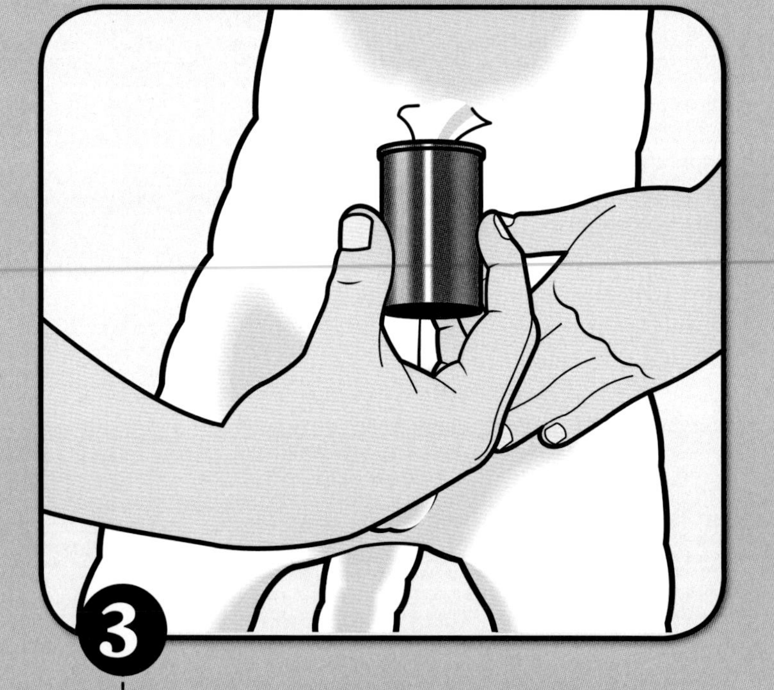

3. Hold the lamb so that the umbilical cord is submerged in the iodine solution. Press the container against the lamb's body.

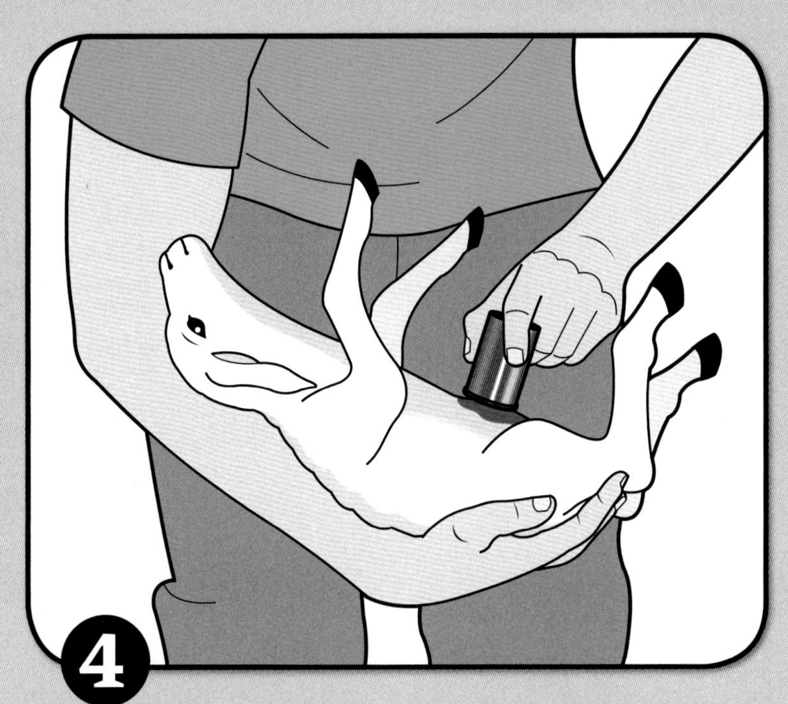

4

Turn the lamb up so that the entire cord and the surrounding area are covered.

## MARKING LAMBS

- A brightly colored, small nylon dog collar or a collar made of yarn is a convenient way to flag any lambs that need special observation; these collars make them stand out in the flock.
- Wax crayons also work well for identifying lambs that need special attention.
- If you have more than two or three ewes, which should produce two to six lambs, the lambs are best identified by ear tags. This way, you can keep records of lamb parentage, date of birth, and growth, and it will be easier to decide which sheep to keep for your flock and which to sell. With identification tags also on your ewes, you can be certain which lambs are whose, even after they are weaned.
- Some tags are a self-clinching type; others need a hole to be punched.
- Tags should be applied while lambs are still penned with their ewes.
- Never use large, heavy cow tags on adult sheep or tags intended for mature sheep on lambs.
- If you're using the small metal lamb tag, insert it into the ear approximately half the length of the tag, to leave growing room for the maturing ear.

# Resuscitating a Lamb

**IF THE HEART IS BEATING** but the lamb is still not breathing, you must swing the lamb to remove excess mucus in the throat and lungs and get it breathing again.

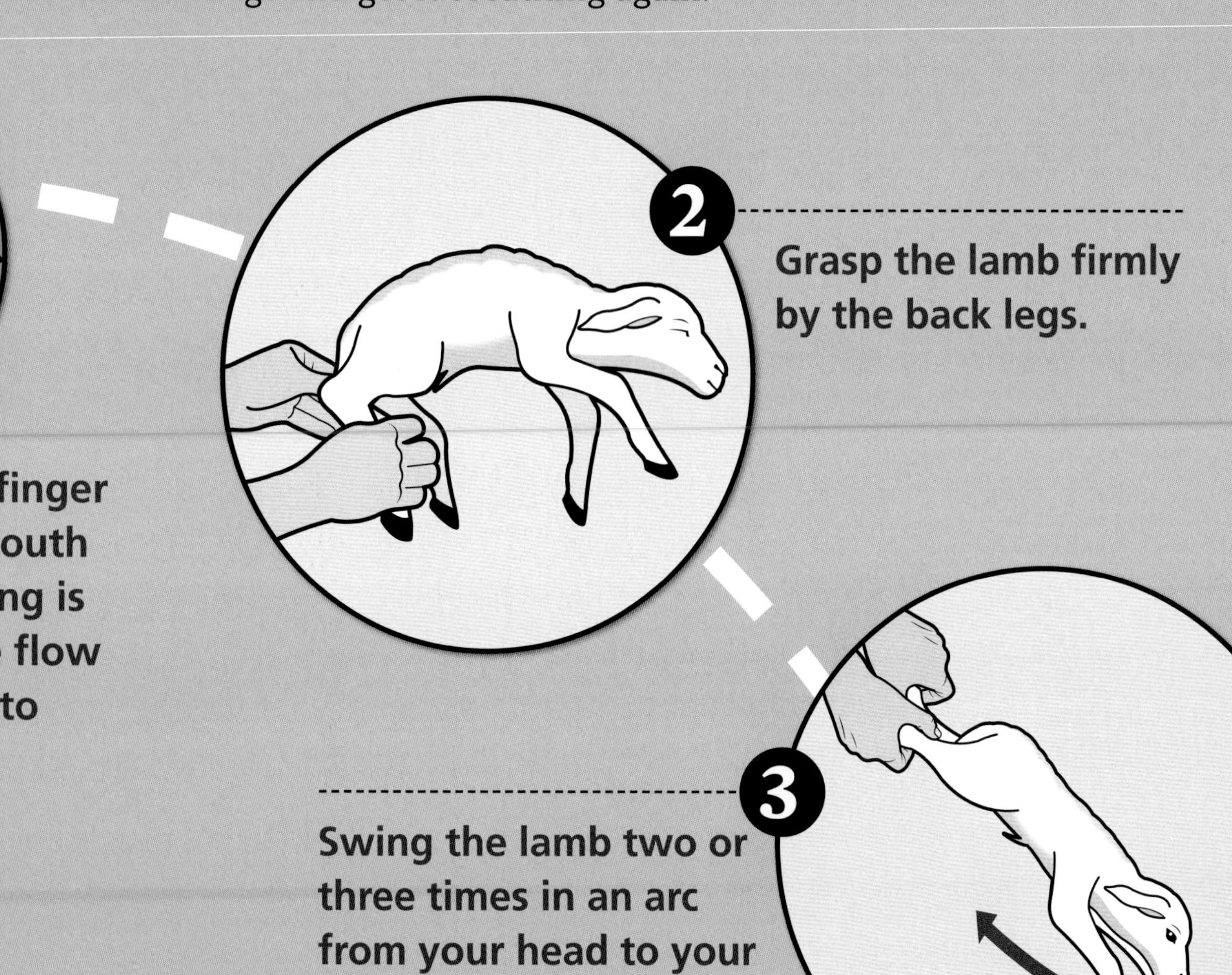

## OTHER TECHNIQUES

Sometimes a cold-water shock treatment will get a lamb breathing again. Dunk the lamb in cold water, such as in a drinking trough. The shock may cause the lamb to gasp and start to breathe. Sometimes a finger inserted gently down the throat will stimulate the coughing reflex.

**4** Place the lamb on its back.

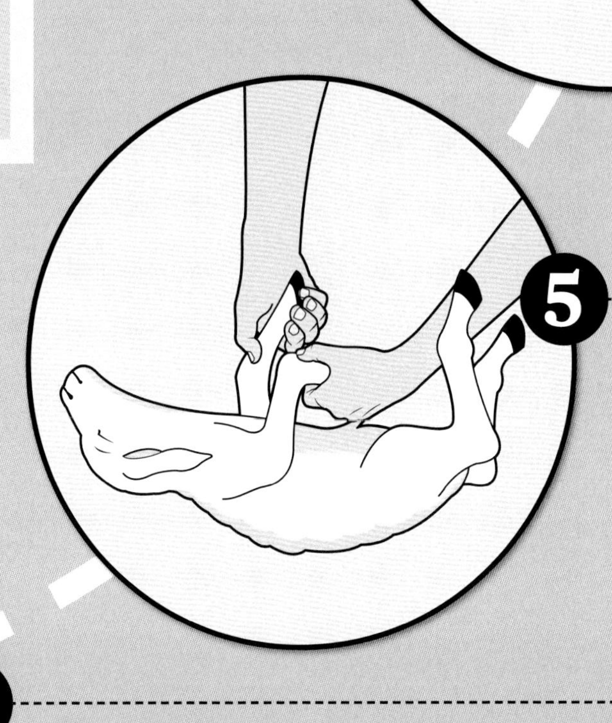

**5** Hold one front leg in each hand and move the legs back and forth over the abdomen.

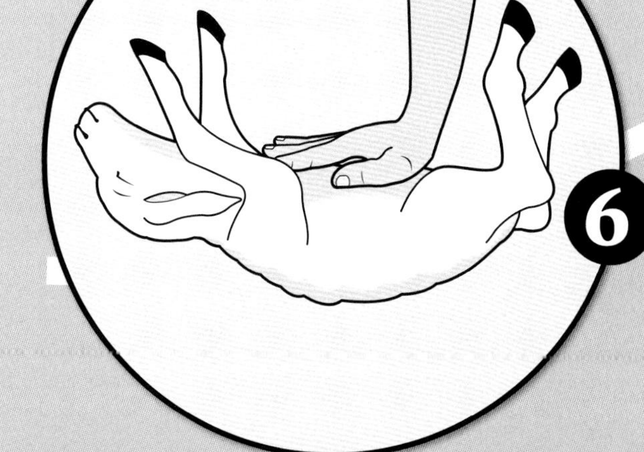

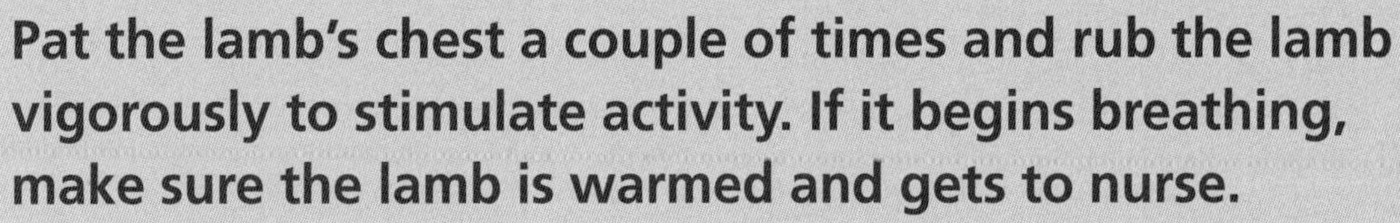

**6** Pat the lamb's chest a couple of times and rub the lamb vigorously to stimulate activity. If it begins breathing, make sure the lamb is warmed and gets to nurse.

# INSERTING STOMACH TUBE

**THE STOMACH TUBE IS A LAST-DITCH EFFORT** to save a lamb's life. Once the lamb's belly is full of milk and it is out of immediate danger, you may need to orphan it. It is easier for two people to feed the lamb with the stomach tube, but one person can do it if the syringe is filled in advance. Prepare by filling the syringe and boiling the tube for 5 to 10 minutes. Let it cool to a comfortably warm temperature.

On a table, hold the lamb's body with your left forearm and with its feet toward you. The lamb's head, neck, and back should be in a straight line but the head should be at a 90-degree angle to the neck.

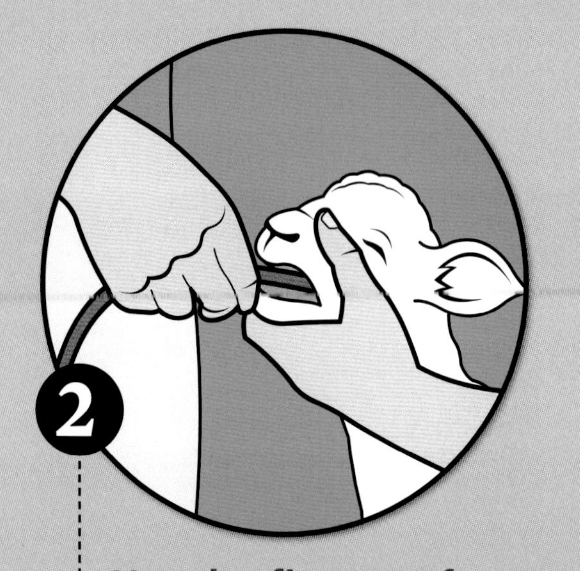


Use the fingers of your left hand to open the lamb's mouth to insert the warm, sterile tube. Insert the tube slowly over the lamb's tongue, into its throat, giving it time to swallow.

Put your thumb and fore-finger along the **LEFT** side of the neck and pass the tube down with your other hand. If you can feel the tube pass, it is going into the stomach.

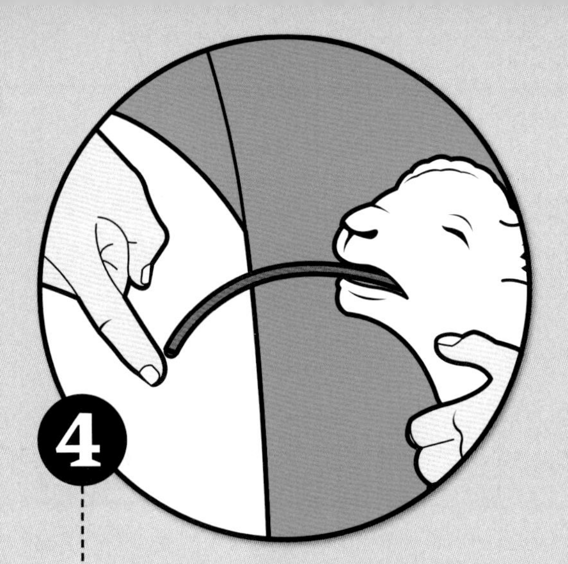


If you think the tube isn't in the correct position, hold a wet finger at the protruding end. If the finger feels cool from moving air, the tube is in the lungs. Remove the tube and start again.

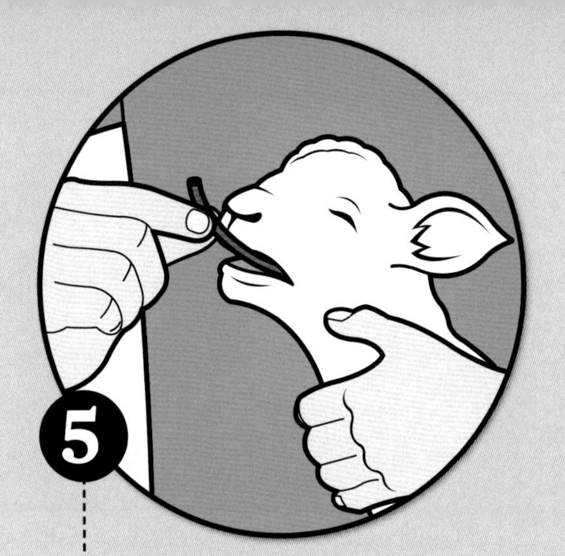


Stop pushing when the end is in the stomach. The average insertion distance is 11 or 12 inches (28 or 30 centimeters). (You cannot insert it too far, but you must insert it far enough.)

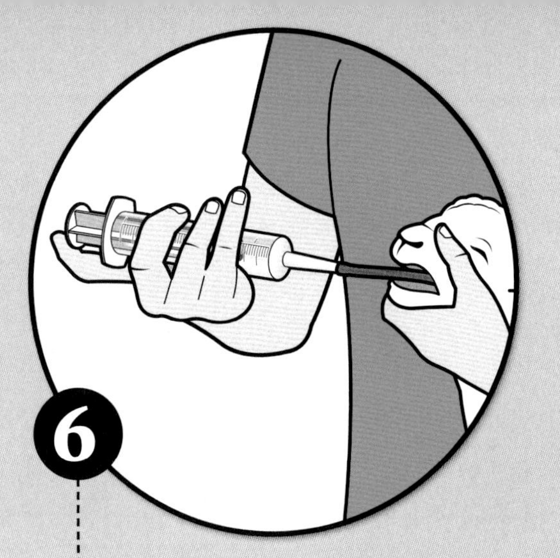


Insert the end of the catheter tube into the syringe filled with warmed milk. Slowly squeeze the milk into the lamb's stomach.

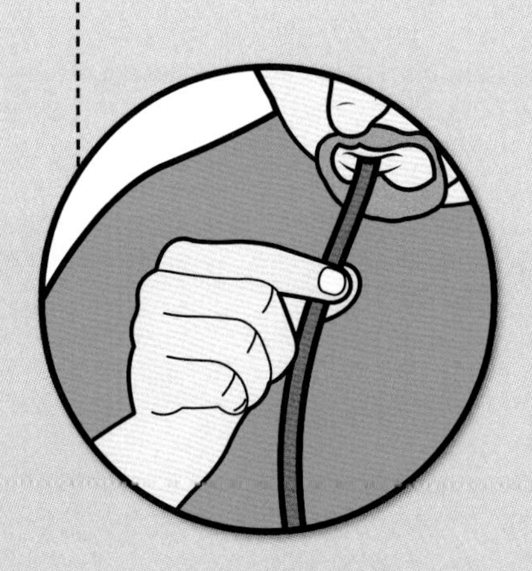

*Another way to check is to blow gently on the tube. If the tube is in the stomach, the abdomen will expand and contract like a balloon.*

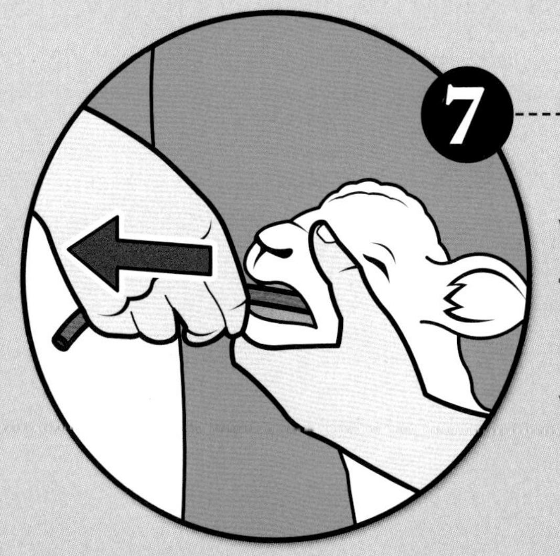

7

Pinch the tube and withdraw it quickly to prevent dripping into the lungs on the way out.

# ORPHANED LAMB

## GRAFTING AN ORPHAN ON A NEW EWE

A ewe may reject one or all of her lambs for a number of reasons. If another ewe goes into labor and you think she may deliver only one lamb, you might choose to graft the rejected lamb on that ewe. Keep the rejected lamb nearby and watch lambing, then follow these steps:

1. Have ready a bucket of warm water and an empty bucket. If you catch the water bag, put its contents into the empty bucket.
2. As the ewe delivers her own lamb, dunk the rejected lamb into the water-bag liquid (if you caught the water bag) or into the warm water up to its head.
3. Rub the two lambs together, especially the tops of the heads and the rear ends.
4. Present them both to the ewe's nose; ideally, she will lick them and claim them both. If she doesn't, wipe them off yourself.

### GRAFTING TIPS

- If the foster mother delivers twins, you may have to dry off the orphan lamb and try again to get its own mother to take it.
- If the orphan is more than 1 week old, you may need to hobble its legs to prevent it from drinking all of the newborn's milk.

## NURSING TIPS

If a ewe rejects her lamb after it starts to nurse, not before, try the following to get her to accept the lamb:

- Check the udder for sensitivity and check the lamb's teeth. A little filing with an emery board can remedy sharp teeth. Don't file too much or the teeth will become sore and the lamb won't nurse.
- Apply Bag Balm to the ewe's teats if they are sore or lacerated by sharp teeth.
- Keep the ewe tied where the lamb can nurse until she accepts it.

# GIVING AN ORPHAN TO A EWE THAT HAS LOST HER LAMB

When a ewe delivers a dead lamb and you have a young orphan that needs a mother, follow these steps. Continue to observe the ewe and lamb carefully to make sure the graft worked. The ewe will be able to smell her own odor from her milk in the lamb's manure, but it takes 3 days for that to happen.

1. **Dunk the lamb in warm water containing a little bit of salt and some molasses.**
2. **Dip your hand in the warm water and wet the lamb's head. By the time the ewe licks off the salt and molasses, she may have adopted the lamb.**
3. **You may want to rub a damp towel over the dead lamb and then rub the towel on the orphan.**
4. **Skin the dead lamb and place it on the live lamb like a sweater for several days, until the ewe has adopted the lamb.**

# Feeding Schedule

**GRAFTING LAMBS TO EWES IS HIT OR MISS,** even when you have a lactating ewe that could be a possible foster mother. Although it is preferable to have lambs nurse ewes, sometimes grafting isn't a possibility and lambs must be orphaned and fed by hand until weaning.

It is important to control the amount of milk that bottle lambs consume during each feeding. A yellow, semi-pasty diarrhea is the first sign of overfeeding. If this occurs, substitute plain water or an oral electrolyte solution, such as Gatorade, for one feeding.

## SUGGESTED FEEDING AMOUNTS

| AGE | AMOUNT |
|---|---|
| 1–2 days | 6 ounces (175 mL) four times a day |
| 3–4 days | 8 ounces (235 mL) three times a day |
| 5–7 days | 10 ounces (295 mL) three times a day |
| 8–14 days | 13 ounces (385 mL) three times a day |
| 15–21 days | up to 16 ounces (475 mL) three times a day; let the lamb decide if it wants to drink the full amount offered |

## HOT FLASHES

After bottle-feeding the lamb, you may notice that it feels very hot or flushed 5 to 10 minutes later. This hot flash usually lasts only a minute or so. Do not become alarmed. It is a known physiologic phenomenon of sheep.

# BUILDING A WARMING BOX

**GUARD NEWBORNS AND YOUNG LAMBS** against hypothermia, which is implicated in about half of all lamb deaths. If you plan to lamb in winter, consider buying, or building, a lamb-warming box.

## WARMING-BOX SPECIFICS

- The mesh floor is elevated in the box about 15 inches (37.5 cm) above the ground, and the box height is 30 inches (75 cm).
- Side vents can be opened or closed to control the temperature generated by a portable heater set up at the hole in the front.
- The lid has one side that's hinged to open and a piece of plastic on the other side for easy viewing.
- If you need something quickly, a box made out of plywood, accompanied by a heat lamp, will do in an emergency.

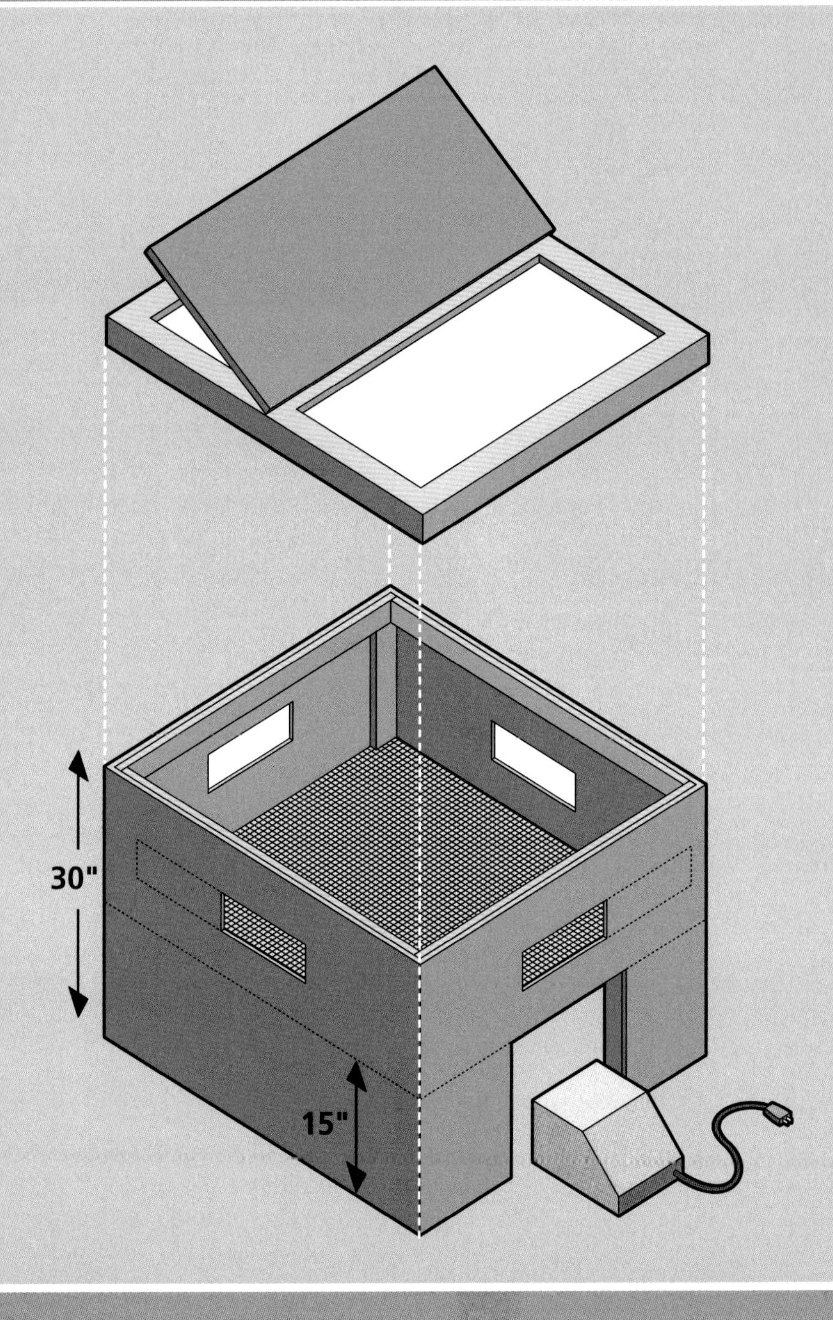


**NOTE:**

A hypothermic lamb will appear stiff and be unable to rise. Its tongue and mouth will feel cold to the touch. You must warm it immediately with an outside heat source.

CHAPTER FIVE

# Wool Production

# SHEARING SUGGESTIONS

**1 SHEAR AS EARLY AS WEATHER PERMITS,** so shearing nicks will heal before fly season.

**2 YOU MAY GENTLY SHEAR EWES BEFORE LAMBING;** this makes it easier to assist them if necessary.

**3 NEVER SHEAR WHEN THE WOOL IS WET OR DAMP.** Damp wool is combustible and can mildew.

**4 PEN THE SHEEP THE AFTERNOON PRIOR TO SHEARING** so they will not be full of feed when you work on them.

**5 SHEAR ON A CLEAN TARP.**

**6 SHEAR BLACK SHEEP & WHITE SHEEP SEPARATELY**, sweeping the floor clean between each.

**7 AVOID SHEARING THE SAME PLACE TWICE.**

**8 SHEAR FLEECE IN ONE PIECE,** but don't trim the wool from the legs or the hooves onto the fleece.

**9 REMOVE DUNG TAGS**, and do not tie them in with the fleece.

**10 SKIRT THE FLEECE** (that is, remove a strip about 3 inches [7.6 cm] wide from the edges of the shorn fleece), especially if you're selling to spinners.

**11 LOCATE YOUR NEAREST SHEARING SCHOOL.** The American Sheep Industry (ASI) Association organizes schools across the country.

# SHEARING

**SHEARING IS A MAJOR JOB** that has to be done each year on most breeds of sheep. You can either hire a professional to do the job or do it yourself. The real "trick" in shearing isn't learning the shearing strokes, but rather immobilizing sheep by the following holds.

## WOOL DENSITY AND FINENESS

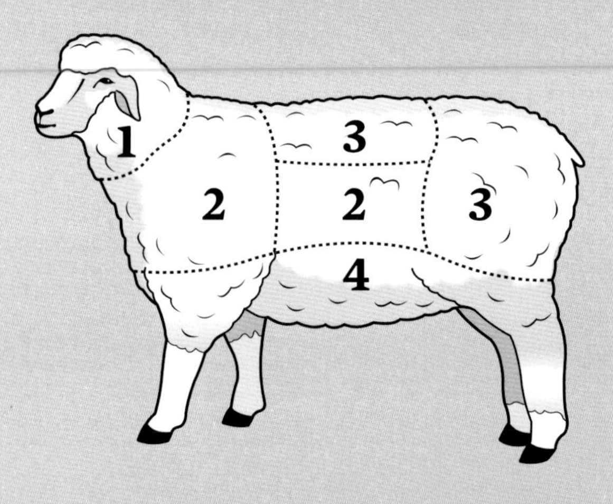


Number 1 tends to be fine and dense wool, and 4 is coarse and thin; 2 and 3 fall somewhere in between.

## 1 BRISKET

Support sheep between knees, with its foreleg firmly on your left side. Make the first stroke **A** straight down the right side of the brisket to the open flank area. Make the second stroke **B** left of brisket, under front leg, and continue down the flank.

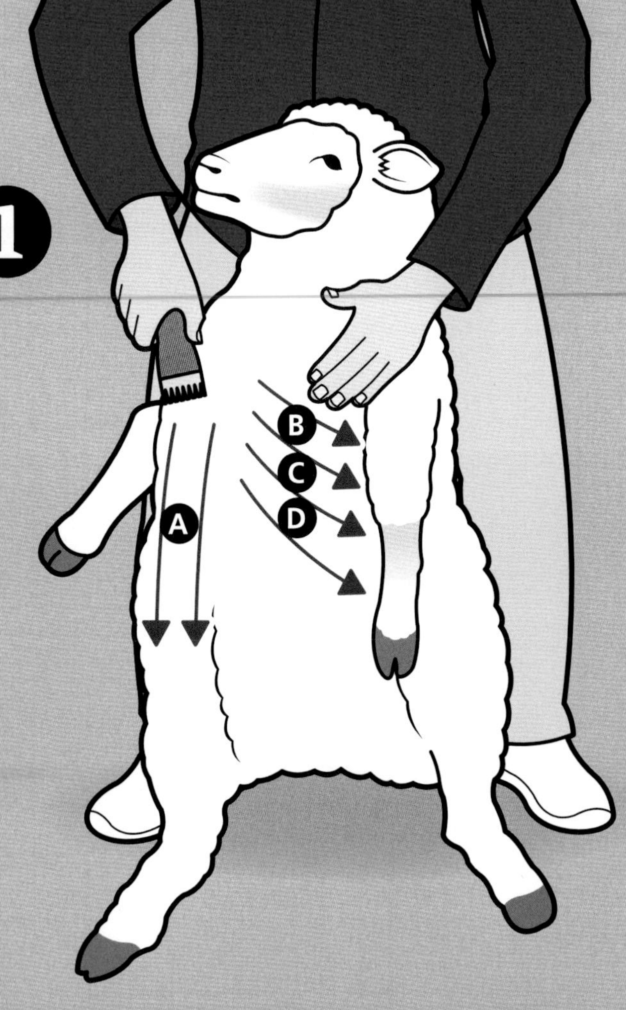

## 2 BELLY

Hold sheep's left front leg with your left wrist and use left hand to tighten skin while shearing belly.

## 3 RIGHT LEG & AROUND UDDER

Stroking downward, shear inside of right back leg A; then come back up leg and around udder.

# SHEARING (CONTINUED)

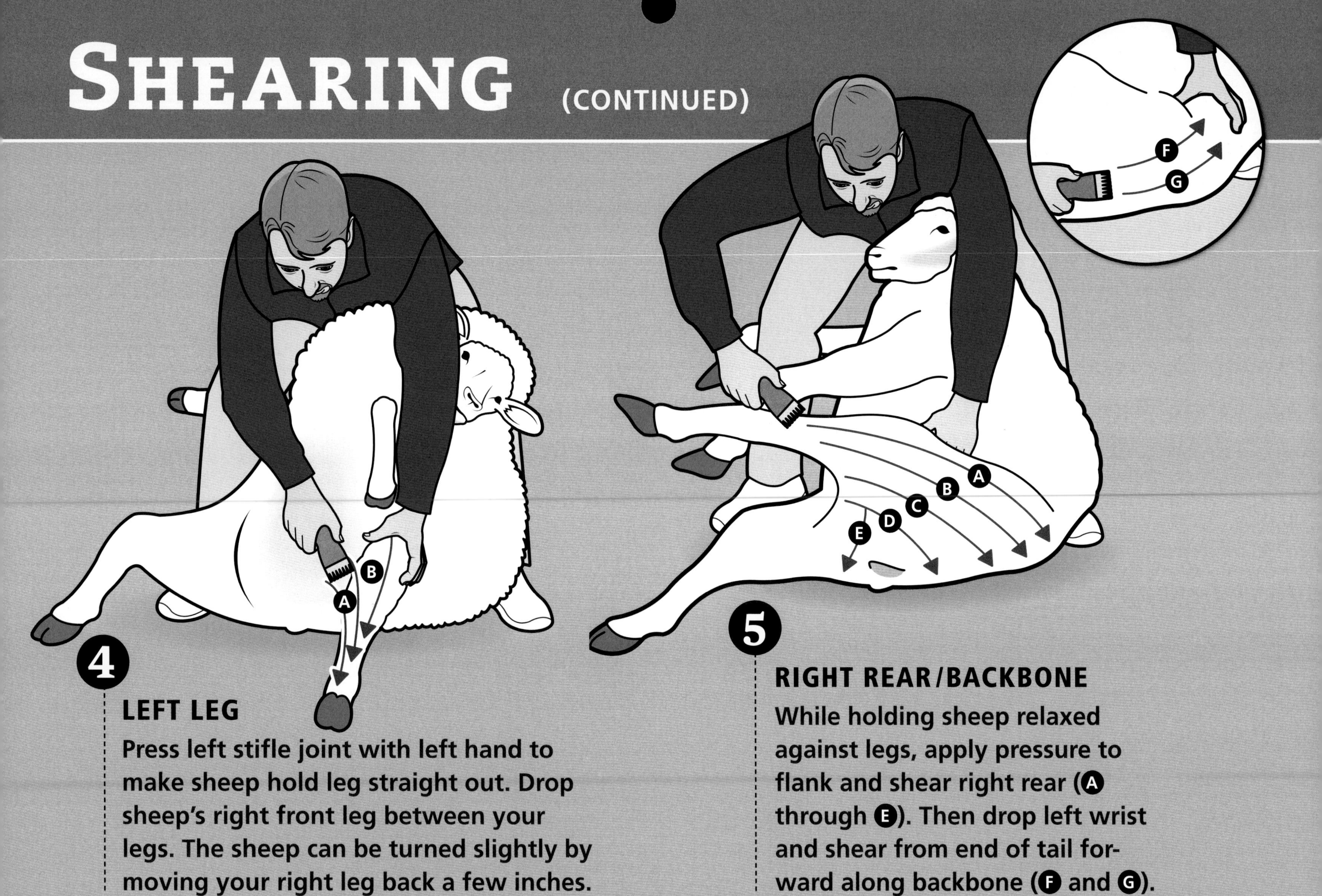


## 4 LEFT LEG

Press left stifle joint with left hand to make sheep hold leg straight out. Drop sheep's right front leg between your legs. The sheep can be turned slightly by moving your right leg back a few inches.

## 5 RIGHT REAR/BACKBONE

While holding sheep relaxed against legs, apply pressure to flank and shear right rear (A through E). Then drop left wrist and shear from end of tail forward along backbone (F and G).

## TOPKNOT

Shift your left hand to sheep's head to shear topknot.

## LENGTH VS. DENSITY

Short wool tends to be denser than long wool; the short wool around the neck is very dense.

## 7

## NECK & EAR

Straighten sheep by placing your left foot close to the sheep's hips and your right foot between sheep's legs. Shear up neck (A and B), turning as you reach the chin so the nose is along your right leg. Shear along neck and jaw (C and D) while shifting left hand toward nose. Finish neck, then hold head against leg with the heel of your hand, grasp ear, and shear it.

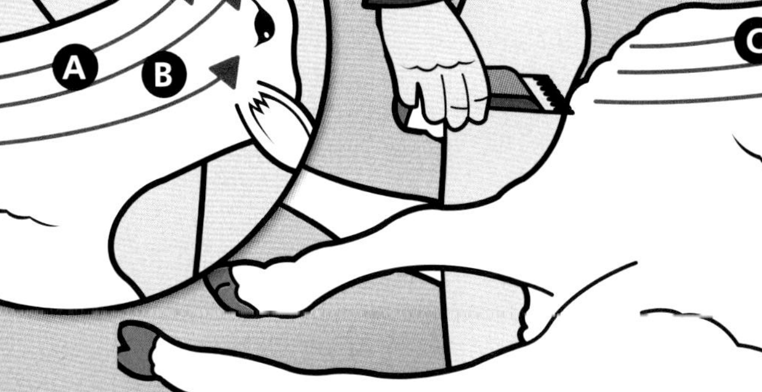

# SHEARING (CONTINUED)

## 8

### UP FRONT LEGS

Hold sheep's body between your knees. Use elbow to hold head against your left leg. Use left hand to stretch skin on shoulder and stroke at knee, shearing upward on front of shoulder Ⓐ. Then take lower part off front legs with downward strokes Ⓑ.

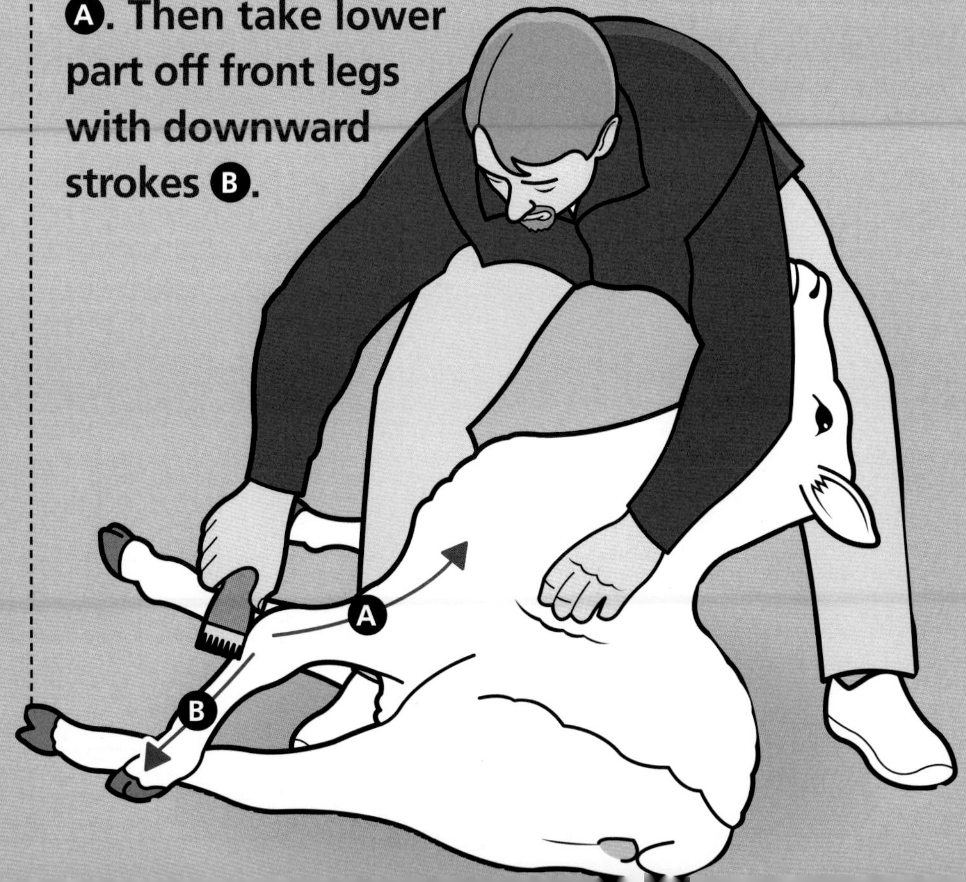


## 9

### DOWN FRONT LEGS

Hold sheep's leg close to your body while shearing down to shoulder.

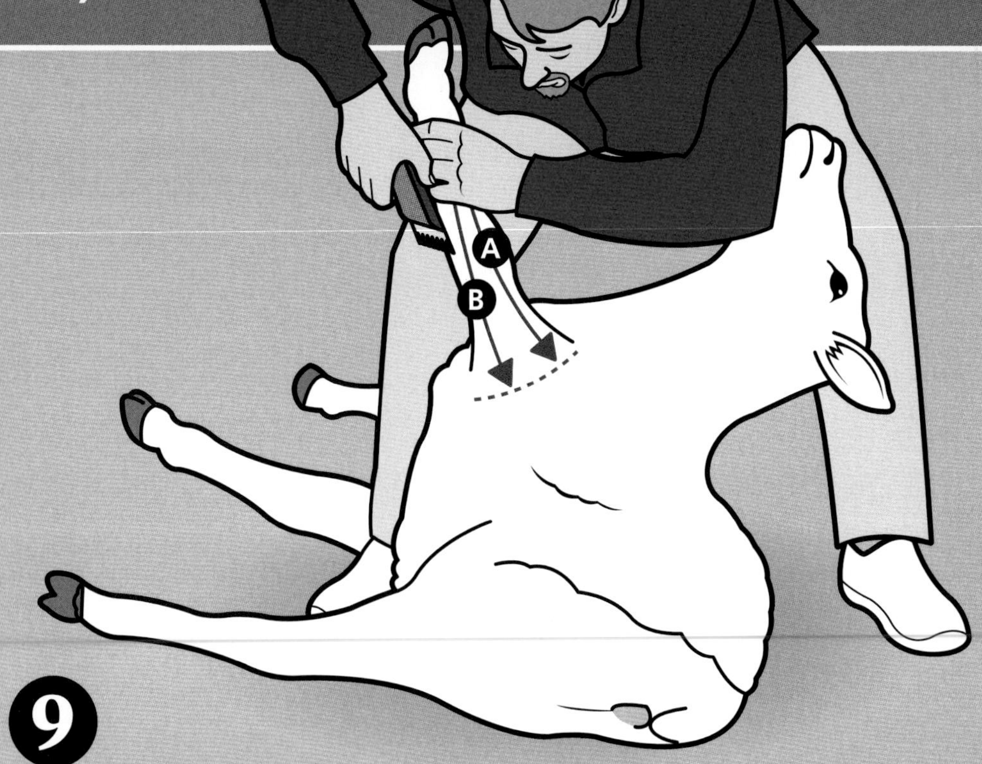


**FACTORS AFFECTING WOOL QUALITY**

Wool quality is affected by genetics, the cleanliness of the fleece, the quality of the shearing job, and the general health of the sheep.

## 10

### BACK

Lay sheep on its back while keeping your right foot between sheep's hind legs and your left foot under its right shoulder. Force left foreleg toward sheep's head to stretch skin on side. Shear along back.

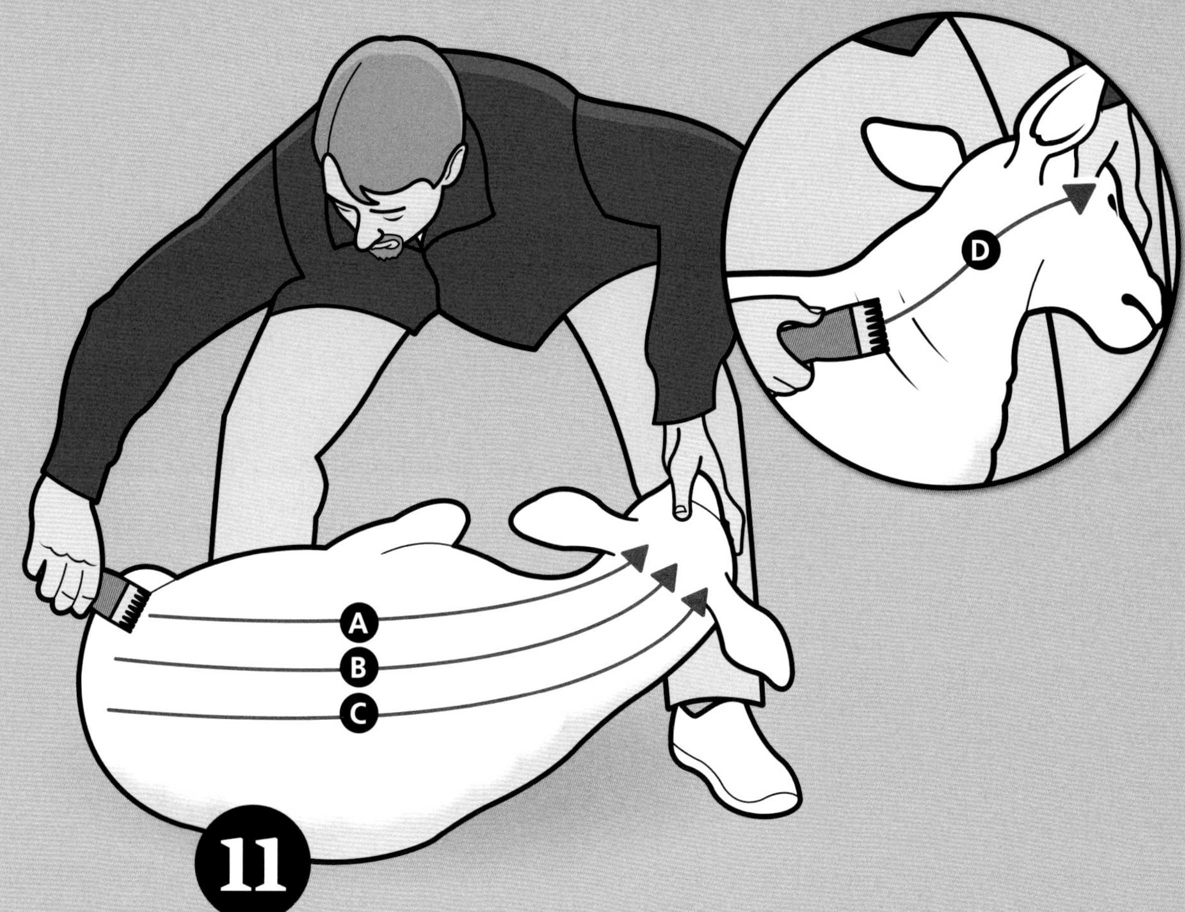


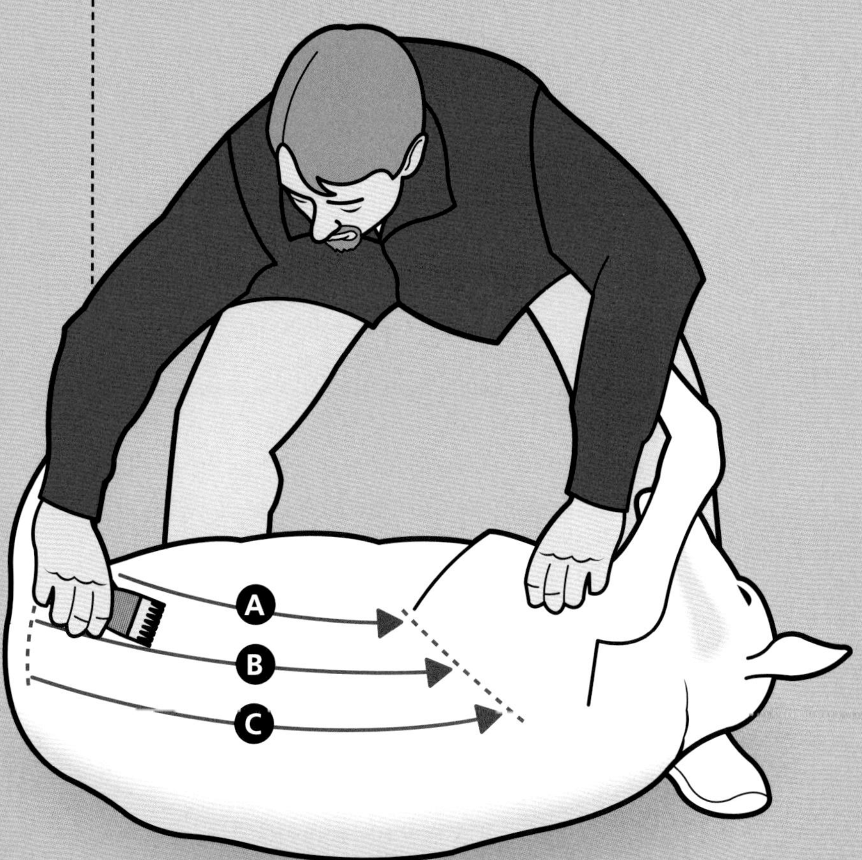


## 11

### BACK (CONTINUED)

Using leg leverage (tail behind your right foot; your left leg under shoulder), roll sheep partly up. Shear along backbone as well as two strokes on lower right side (A through C). Holding up right ear with left hand, do final stroke just over backbone and go all the way under ear to jaw D.

# SHEARING (CONTINUED)

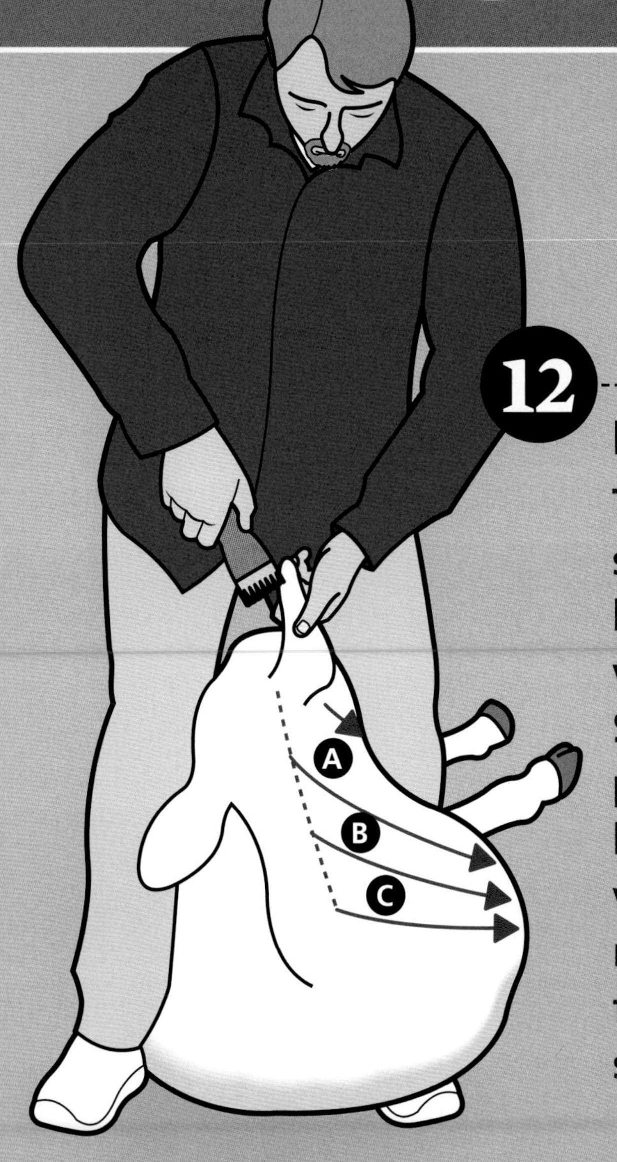


## 12 EAR & FACE

Turn head up. Clasp sheep between your legs and finish clearing wool from ear and face. Shear straight down to point of shoulder using left hand to straighten wrinkles. Allow sheep's right foreleg to come forward after the third stroke.

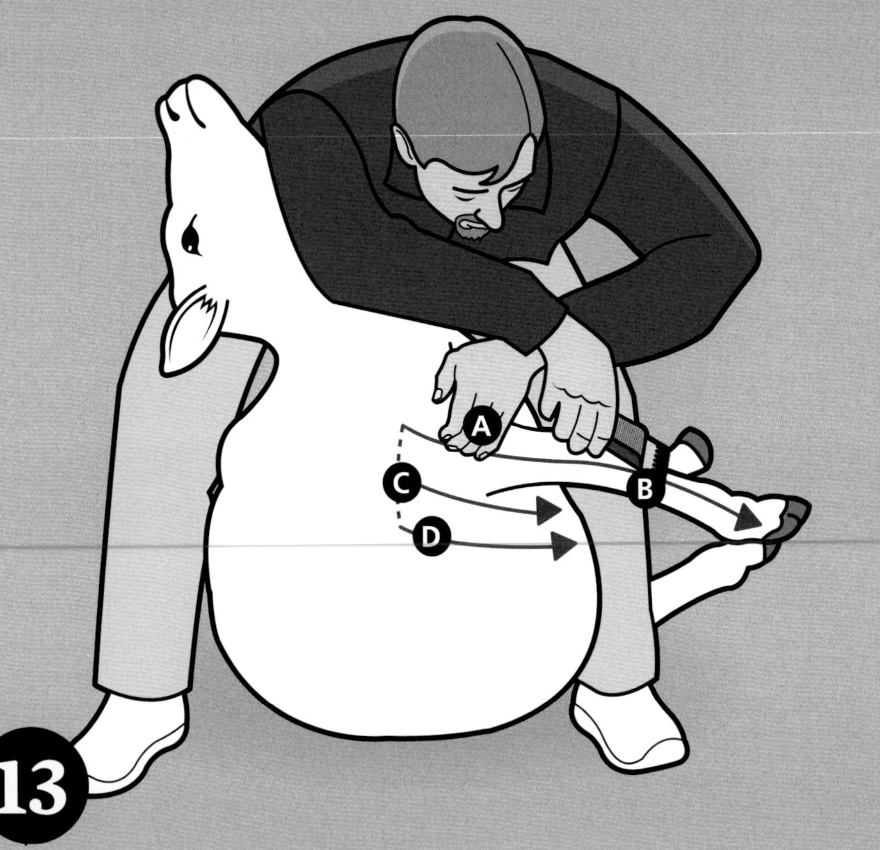


## 13 UNDER FRONT LEG

While still straddling sheep, push down on point of shoulder with palm of hand while pulling up skin with fingers to clean under front leg. Final stroke should continue under front leg to toe.

## 14

### HIND END & LEGS

At last stroke, move left foot to other side of sheep's hind legs and raise head to continue down this far side.

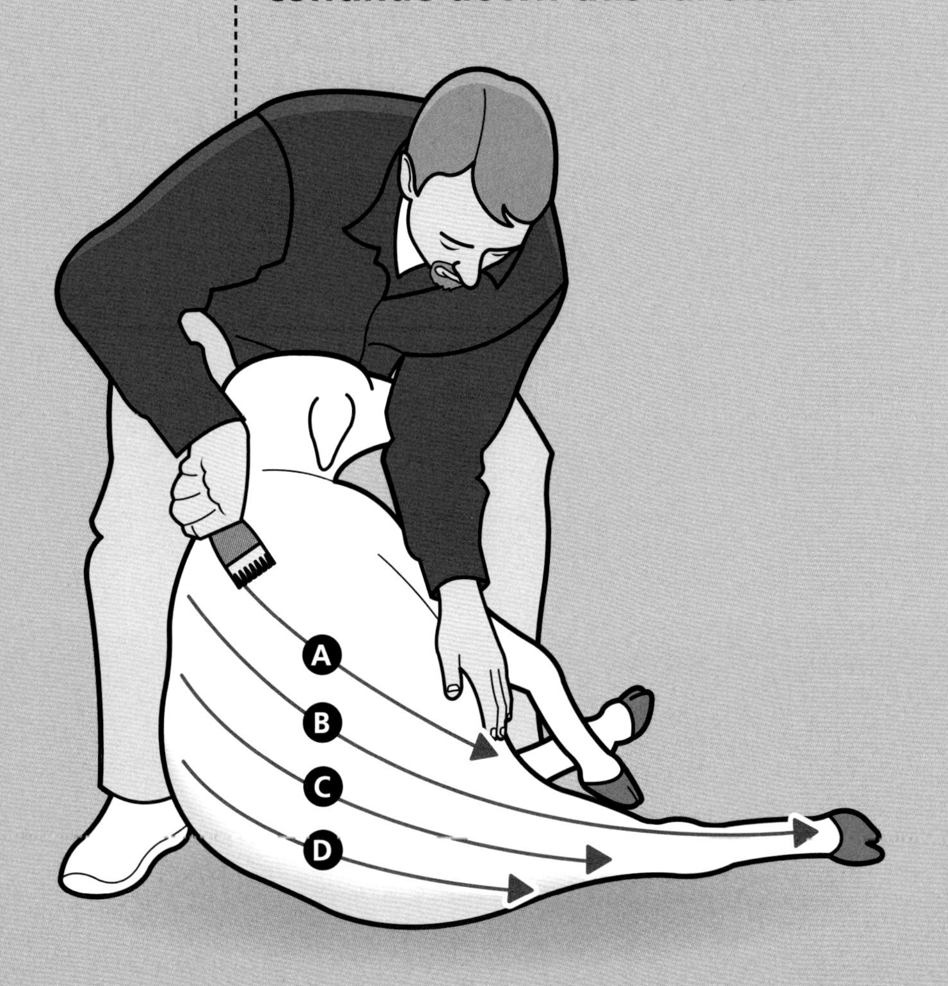


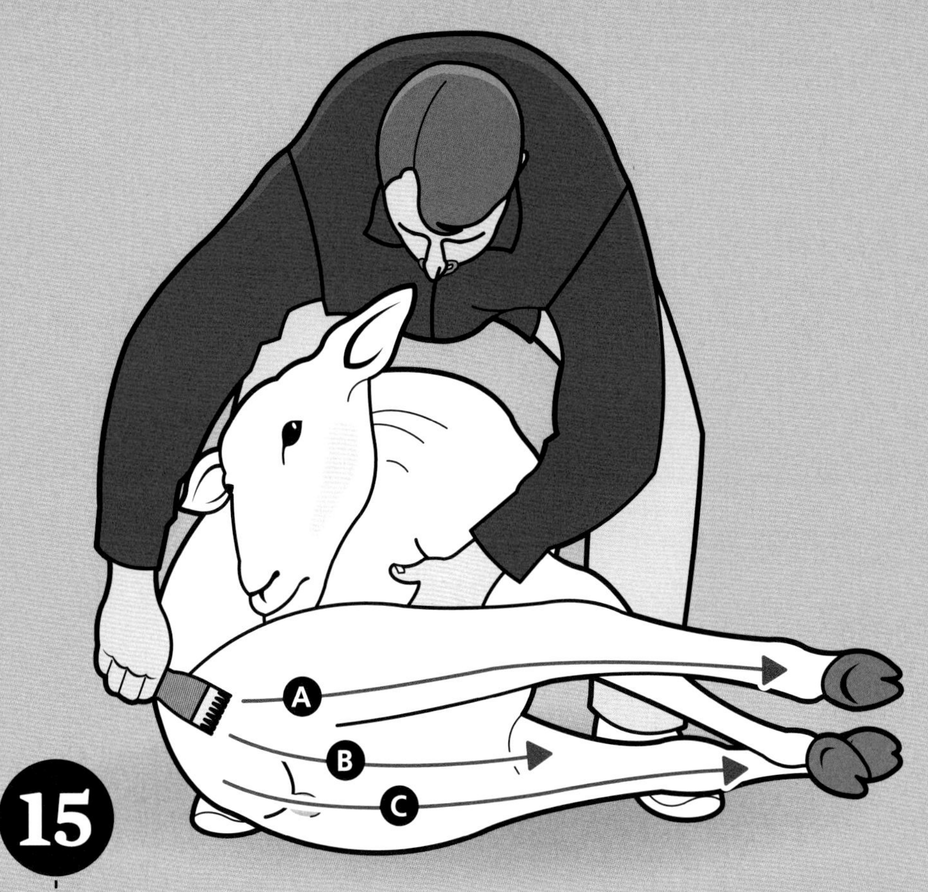


## 15

### HIND LEGS

With sheep in position, hold skin of right flank in left hand and apply pressure to stifle joint. Hold sheep's head between left arm and right leg so sheep settles down on its side. Move left foot back as sheep settles and right foot forward to allow you to reach the last strokes. The left foot keeps the sheep's shoulder and feet off the floor.

# HAND-CARDING

**THE CARDING PROCESS "TEASES" APART THE FIBERS,** removes short fibers, and sets the fibers to lie in the same direction. During carding, you can blend various fibers, yielding interesting colors and textures. Hand-cards work well for occasional small carding jobs, and they are not very expensive if you're just starting out. Be sure to pick out any large pieces of vegetation before you begin carding.

### READY TO SPIN

Once your wool is prepared, it can be spun into yarn or used in other handcraft projects.

**1** Spread the wool on the left hand-card, with the shorn ends at the top of the card.

**2** Take the right hand-card and lay it in the center of the left hand-card, with the handles in opposite directions, and draw the right hand-card away from you. Repeat several times until fibers begin to align themselves.

## DRUM CARDER

If you have to card the wool from several sheep, you may want to try using a hand-operated drum carder. It's faster and more efficient, and will save your wrists from carpal tunnel syndrome.

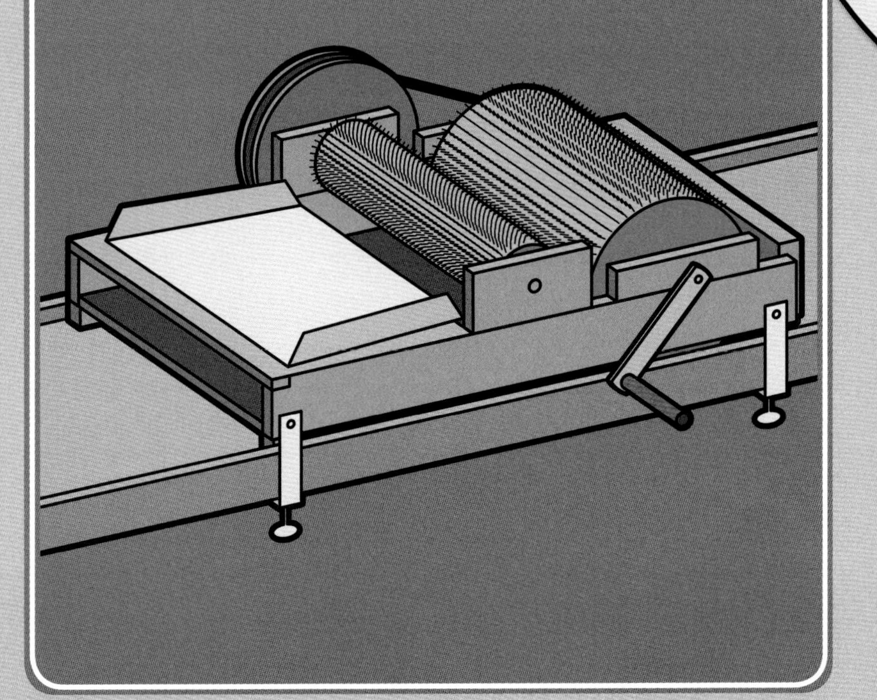

**3** When the fibers are well aligned, lay the right hand-card on your knee and with the handles in the same direction, brush away from yourself. This deposits the wool on the right hand-card. Switch paddles and repeat this step several times.

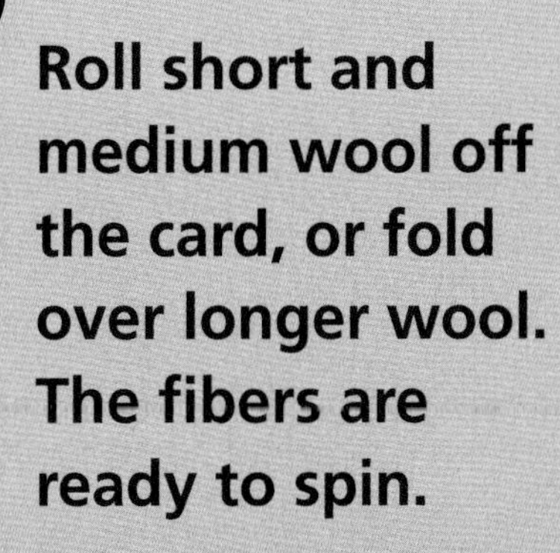

**4** Roll short and medium wool off the card, or fold over longer wool. The fibers are ready to spin.

# Fiber Structure

**WOOL IS PRODUCED BY FOLLICLES,** which are cells that are located in the skin. There are two types of follicles: primary and secondary. Primary follicles can produce three kinds of fiber: true wool fibers, medulated fibers, and kemp fibers. Secondary follicles produce only true wool fibers. The medulated fibers are hairlike fibers that are as long as the true wool fibers but lack the elasticity and crimp (or waviness) of true wool. The kemp fibers are coarse and typically shed out with the seasons. Kemp fibers don't take dyes well, but kemp is important in the production of certain types of fabrics, like true tweeds, and in the production of carpet wool.

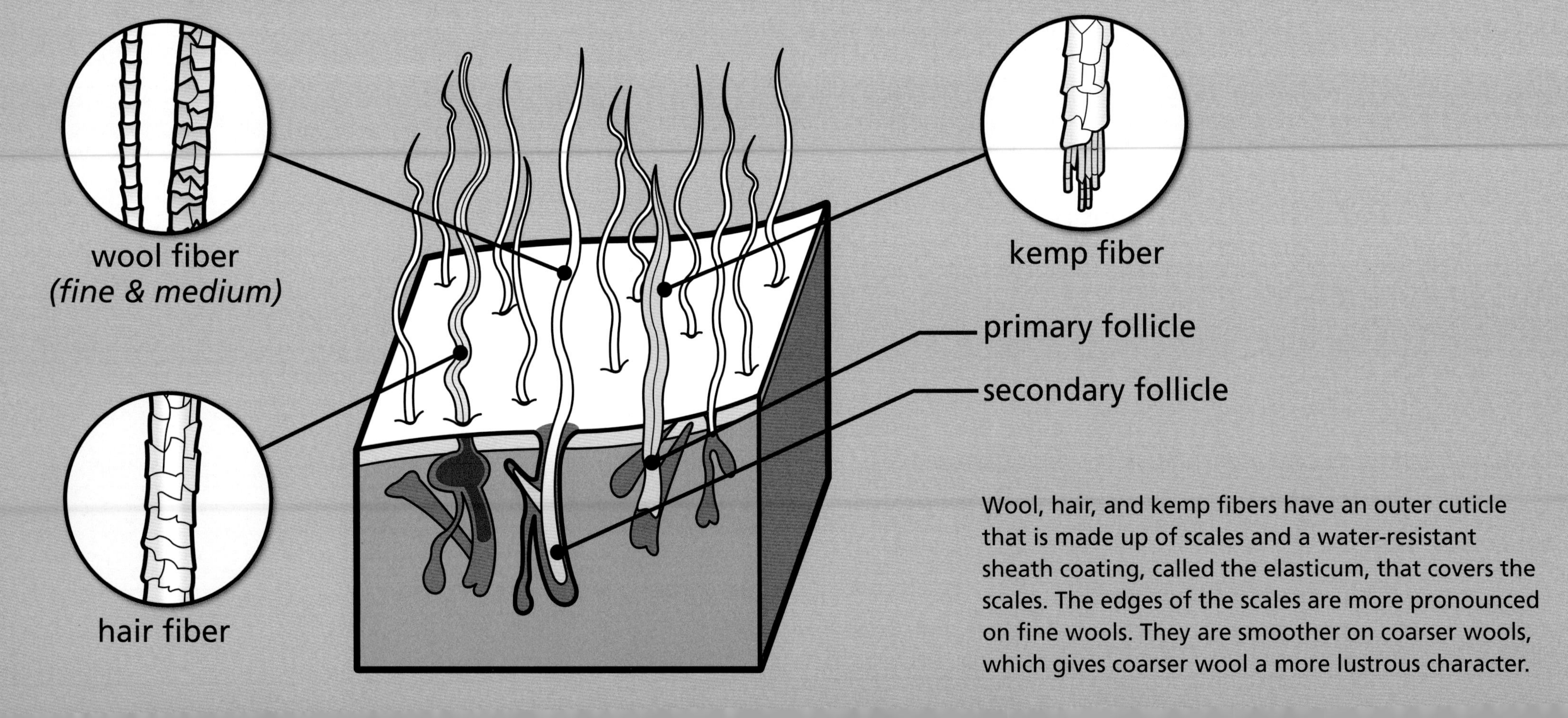


Wool, hair, and kemp fibers have an outer cuticle that is made up of scales and a water-resistant sheath coating, called the elasticum, that covers the scales. The edges of the scales are more pronounced on fine wools. They are smoother on coarser wools, which gives coarser wool a more lustrous character.

# EVALUATING WOOL

**COMMERCIAL BUYERS PURCHASE WOOL** by the pound (0.5 kg), which is put in a bag, or bale, that can weigh anywhere from 150 to 1,000 pounds (75–500 kg). These buyers evaluate the clip, or the season's yield of wool, from a flock or group of flocks on the basis of several characteristics, depending on the wool's intended use. The following are some of the criteria that a wool buyer considers.

## Yield

Yield is calculated on the basis of a sample that's weighed raw, scoured and dried, and reweighed. The average in North America is 45–65 percent.

## Quantity & Type of Vegetable Matter

A wool clip that has lots of vegetable matter (seeds, twigs, etc.) is much less valuable than one that doesn't contain any.

## Average Length & Variability of Staple

The length of fibers falls into three major classes: staple, French combing, and clothing. Buyers want minimal variability within a fleece.

## Crimp

Again, buyers are interested in consistent quality.

## Staple Strength & Position of Break

Tender wool has low tensile strength and breaks unevenly, whereas broken wool breaks at the same point on most fibers throughout the fleece.

## Color & Colored Fibers

Most commercial buyers want mainly bright, white wool that can be dyed without bleaching; many hand-spinners want naturally colored wools.

## Fiber Diameter

"Spinning count," "blood grading," and "micron" systems are used to describe fiber diameter.

CHAPTER SIX

# Record Keeping

# MARKETING CONSIDERATIONS

**BELOW ARE SOME MARKETING POSSIBILITIES,** with considerations both positive and negative listed for each type of possibility. It is important to fully consider your marketing options before diving in.

## FREEZER LAMB

- Carries a premium price opportunity
- Is conducive to either grass-fed or grain-fed production systems
- Requires initial advertising to develop a market
- Requires organization, time, and maintenance of customer database, order forms, etc.
- Requires people skills to develop and maintain good customer relations
- Requires good timing, communication, and transportation

## CLASSIFIED AD

- Draws customer to you, so no transportation of animals
- Requires alternative plan if there are no buyers
- Requires knowledge of market and ability to haggle to get a better price
- Takes more time to be made available to customers

## BREEDING STOCK

- Requires increased expertise in selection and breeding of top genetics
- Requires increased investment for your start-up flock and new genetic influxes
- Requires increased promotional costs and skills; may require dedicated show ring efforts
- Presents increased profit potential
- Requires additional record keeping

## SALE BARN

- Presents lower profit potential, with some potential to reach market peaks for ethnic and religious holidays
- Eases dispersal of goods if location and transportation are convenient
- Requires you to sanitize vehicle and clothes to reduce risk of transporting disease pathogens home to your flock
- Allows you to reach market peaks for ethnic or religious holidays
- Increases stress on animals

# Meat Cuts

**THERE ARE SEVERAL CORRECT WAYS** to break up a lamb carcass, and no one method can be considered best. The technique shown is commonly used.

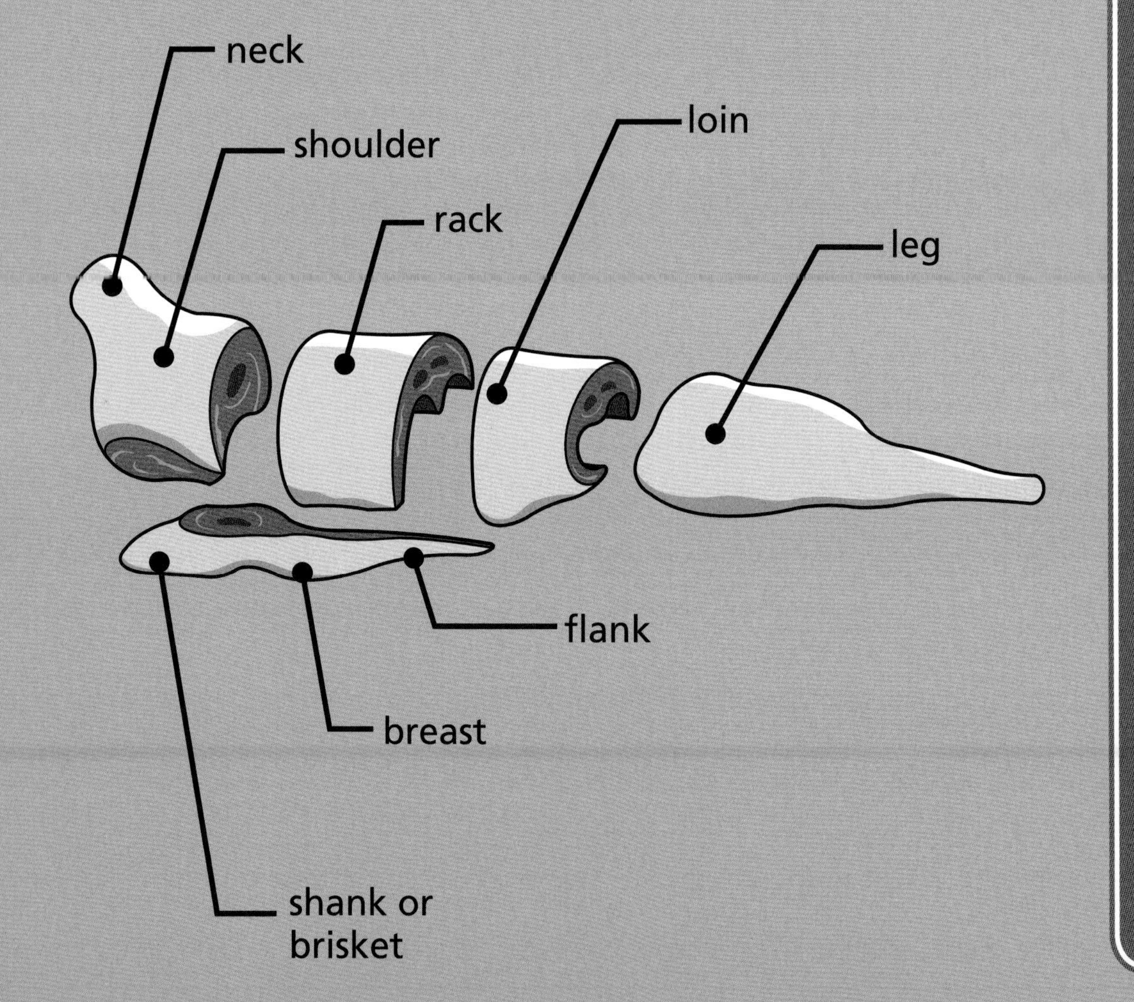


## INSTRUCTIONS FOR PACKAGER

*To get the maximum use and enjoyment from your sheep, make the following requests for packaging:*

- Cut off the lower part of hind legs for soup bones.
- For mutton (sheep older than 1 year), smoke both hind legs for "hams."
- For lambs (sheep younger than 1 year), leave hind legs whole or cut into sirloin roasts or steaks.
- Cut the loin from mutton (or lamb) as tenderloin or as a loin roast.
- Package riblets, spareribs, and breast meat into 2-pound (1–kg) packages. (Riblets need to be well cooked, such as in a pressure cooker for 45 minutes.)
- For mutton, have the rest boned, ground, and the fat trimmed. Double-wrap in 1-lb (0.5-kg) packages.
- Cut the rack, or rib area, of the lamb into "lamb chops" or leave as a rack roast. Cut the shoulder into roasts or chops, (The neck and shank can be used as soup bones. Stew meat, or ground lamb, can also come from these "front" cuts.)

YEAR______ 

# SAMPLE EWE RECORD CHART

| Ewe name or number | Number of live lambs born (DATE) | Average weaning weight* | Number of weaned lambs (DATE) | Wool condition | Wool weight* (DATE) | Vaccinations, hoof trimmings, illnesses, other information (DATE) |
|---|---|---|---|---|---|---|
| *101* | *2 (5/13)* | *48* | *2 (7/19)* | *excellent* | *7 (4/28)* | *trimmed hooves (4/28)* |
| *102* | *2 (5/19)* | *39* | *1 (7/19)* | *excellent* | *7.5 (4/28)* | *trimmed hooves (4/28); 1 lamb died of scours (5/22)* |
| | | | | | | |
| | | | | | | |
| | | | | | | |
| | | | | | | |
| | | | | | | |
| | | | | | | |
| | | | | | | |
| | | | | | | |

*Weight in pounds.

# LAMBING SCHEDULES

**NORMAL GESTATION,** the time between breeding and lambing, is from 147 to 153 days (21 to 22 weeks). Use the chart below to keep track of your ewes' schedules. You may want to begin by filling in the due date and then work backward.

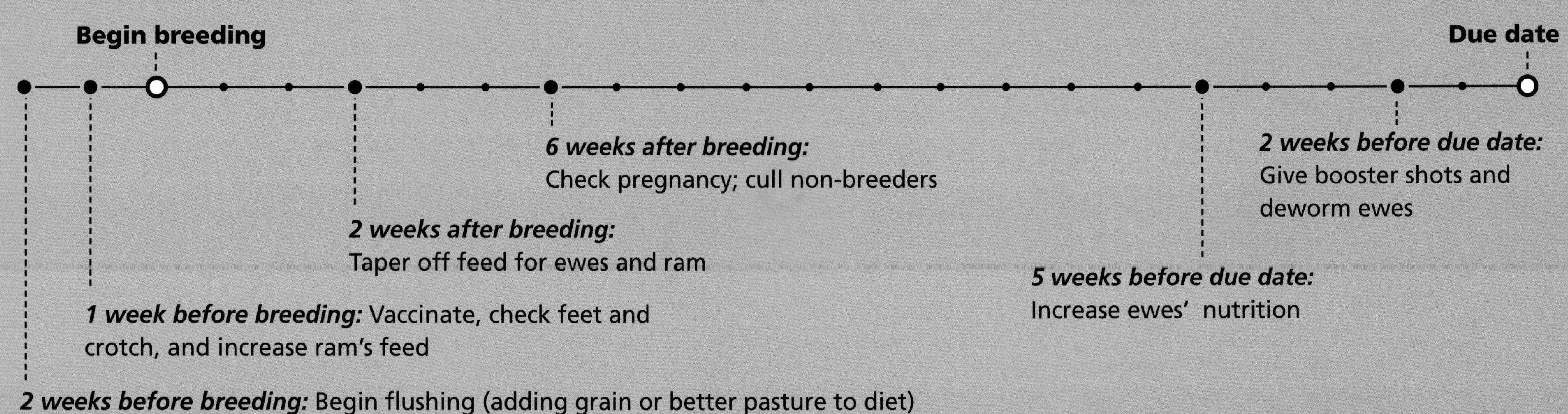


| EWES | Begin flushing | Check feet & crotch, vaccinate | Begin breeding | Taper off feed | Pregnancy check; cull non-breeders | Increase ewes' nutrition | Booster shots & deworming | Desired lambing date |
|---|---|---|---|---|---|---|---|---|
| | | | | | | | | |
| | | | | | | | | |
| | | | | | | | | |
| | | | | | | | | |
| | | | | | | | | |
| | | | | | | | | |